GEOPEKO

A successful Australian mineral explorer

Edited by Brian Williams and Rob Ryan

Editors: Brian Williams, Rob Ryan
Contributors: Andrew Browne, Wendy Corbett, Brett Duck, John Elliston, Geoff Eupene, John Goulevitch, Jeff Gresham, Tony Hope, Gary Jones, Greg Kater, Peter Kitto, Ross Large, Michael Love, Warwick Maehl, Paul McInnes, Bob McNeil, David O'Connor, Wayne O'Neil, Terry Quinlan, Bob Richardson, Geoff Sherrington, Colin Sinclair, Alex Taube, Ray Twist, Kim Wright.

Published in 2018 by Connor Court Publishing Pty Ltd

Copyright © Brian Williams and Rob Ryan (as editors)

All rights reserved. No part of this book may be reproduced or transmitted in any form or by any means, electronic or mechanical, including photocopying, recording or by any information storage and retrieval system, without prior permission in writing from the publisher.

Connor Court Publishing Pty Ltd
PO Box 7257
Redland Bay QLD 4165, Australia
sales@connorcourt.com
www.connorcourt.com
Phone 0497-900-685

ISBN: 9781925501780 (Paperback)
        9781925501797 (Hardback)

Front Cover Design: Maria Giordano

Printed in Australia

PREFACE

Since early man first lit a fire and discovered metal in the ashes the use of mineral resources has underpinned much of mankind's progress. From the steel and concrete that bridges the rivers and supports the high rises in the cities, where half the population now lives, to the rare and, to many people, unheard of elements that lie behind the smart phone and the exotic medicines that keep many people alive; from the electricity that powers the food preservation appliances to the nuclear power stations that supply massive quantities of energy without releasing any carbon dioxide, the resources industry lies behind almost every improvement in the quality of life that humankind has been able to develop over the few thousand years it has been around.

It is the job of the earth scientists to find new resources to replace those that are being consumed at an ever-increasing rate, and they can only be found where nature has put them.

Listening to some of the anti-mining monologue one could be forgiven for thinking that the mining and energy companies exist purely to make a profit for their cigar-smoking owners. 'Big Oil' and 'Big Coal' are deemed to be behind any research that contradicts the ideology of the anti-capitalist movements around the world.

Of late, articles and presentations bemoaning the falling rate of discovery of new mineral and energy resources and the start-up of new production have bombarded the mining industry. As the message reads, this is attributed to intellectual failure by the professionals in the minerals industry to do what the academics and political advisers want. It is implied that if only they would act smarter, and be nicer to the communities where the resources are found, the problems would go away.

The arguments ignore the basic issues that affect the search for, and extraction of, mineral and energy resources. These can be classified as – location, lead times, depletion, discoverability, development issues, and 'boom-and-bust'.

Location

Nature has decided where the resources are, not the explorers. While there are sufficient resources available it may be acceptable to forgo the exploitation of some in the interests of conservation and the environment, but sooner or later the demand for new production will lead to the development of those that had previously been protected.

Lead Times

It takes time and money to find a new deposit of minerals or a new source of energy. Usually several years and several million dollars will have been spent before the point is reached where a deposit has been found and a decision to investigate the probability of developing it can be taken. If that decision is positive, more time and money will then be committed to the study of whether development is feasible. If the conclusion, then, is that the project is financially viable, even more time and money are required for development to the production stage. The process will take years and require the investment of tens, or even hundreds, of millions of dollars.

Depletion

Depletion is self-evident. The world's mineral and energy resources are finite. In due course they will all run out. Even the sun will supernova one day. Minerals that have been produced over the past centuries are either locked up in infrastructure and in manufactured machines and appliances, or dumped somewhere. A proportion can be recovered by reclamation and recycling, but unless we are prepared to knock down all the cities and their infrastructure, and recycle their contents, a substantial proportion of the original mineral resource is now unavailable. And energy, once consumed, cannot be recovered.

Both will have to be won from new sources around the globe and, eventually, from elsewhere in the universe.

Discoverability

A substantial proportion of the easily found near-surface mineral and energy deposits have probably been found in those countries where there has been active exploration in the past – which is most of them. Certainly fewer and fewer are there to be located easily and cheaply. In the future it will be necessary to look deeper into the crust with better technology and an improved understanding of how mineral and energy-rich deposits form, and where, and it will cost more.

Development issues

Mining inevitably disturbs the environment. When populations were small and environmental issues had not yet developed, the societies of the day ignored the impacts that mining had, unless they were very severe, accepting the benefits that flowed. Today the industry is required to, and accepts, that environmentally acceptable mining must be the norm and that operations must be run as cleanly as possible. However, there will always be some impact, and costs in meeting the requirement to prevent or remediate those impacts, which must be passed on.

'Boom-and-bust'

As populations grow and demand a better life-style, and as new technology creates demand for new and different metals, production at the time cannot satisfy the rising demand, prices begin to rise, and another resources boom arrives.

That underwrites the decisions by the investors and explorers to search for and, if successful, develop new supplies. But the lead times involved in that process, and the fact that investors, having committed the funds and the time, press on to production, mean that a few years later there is a rush of production, a glut, and a collapse in prices, putting many of the operations either out of business or on to a 'care-and-maintenance' status pending the next boom.

Very few mineral deposits have a constant grade[1] throughout or are defined by clear-cut boundaries between ore and waste. They usually fall away in grade from a richer centre to a poorer and indeterminate boundary. Therefore, what proportion of an orebody can be produced at any time is determined by the market price of that commodity and the cost of extracting it. As prices fall, either a mine is shut down or 'high-grading' is necessary, meaning that what is left behind is of lower grade and no longer has a higher grade component to support production to the same extent in the next boom. Thus it is less likely that it can be brought

---

[1] Grade – the percentage of metal present in the ore.

back into production when the opportunity arises, unless prices have risen further. And as it becomes necessary to mine deeper to access these resources, so the costs rise.

This all sounds like stating the obvious but seems not to have been grasped by the politicians or the media. It is incumbent upon the industry to tell the story and get people to recognise the facts, in particular the limits to how much energy and minerals are left on the earth.

Maybe people will then stop wasting them.

Although, to an outsider, exploration crews "running around in jeeps" and emerging from the bush dishevelled and dirty may seem to be a bit rough and ready mineral exploration, when practised professionally, is a highly scientific and skilled occupation, as this book hopes to demonstrate. And because of its somewhat esoteric nature, it can appear a bit mysterious to those not trained in it.

It also requires excellent management, attention to the safety of crews who may be asked to work in dangerous locations, concern for those who live where the minerals are found, and attention to cleaning up the site after any mineral or energy deposit has been mined out.

It is the hope of the contributors to this book that readers may put it down with a better understanding of, and appreciation of, an industry that sustains their way of life.

CONTENTS

INTRODUCTION	13
CORPORATE HISTORY OF PEKO-WALLSEND	17
THE FOUNDATIONS OF PEKO AND GEOPEKO	33
PROJECT GENERATION	39
MINE SERVICES	43
OPERATIONS: 1963-2000	53
NORTHERN TERRITORY	54
Tennant Creek	54
Darwin and Jabiru	75
Cooper Creek and Borrodaile	79
South Alligator Joint Venture	82
Burrundie	83
Diamonds	85
QUEENSLAND	85
North-west Queensland	85
Mt Morgan	87
Mt Chalmers	88
Townsville	90
Brisbane	91
Mt Isa	94
TASMANIA	95
King Island	95
Devonport	97
NEW SOUTH WALES	99
Coal Group	99
Parkes	100
Peak Hill	105
Cobar	107

WESTERN AUSTRALIA	109
Perth	109
Paterson Range	109
Peak Hill	109
Bangemall Basin	110
Kalgoorlie	112
The Platinum Saga	115
SOUTH AUSTRALIA	116
Adelaide	116
VICTORIA	118
Ballarat and Ararat	118
OVERSEAS EXPLORATION	123
USA	124
New Caledonia	127
Chile	127
New Zealand	132
Malaysia and Indonesia	132
Europe and Africa	133

INNOVATION AND INVENTION	139
ORE GENESIS	142
THE RICHARDSON CONTRIBUTION	144
THE INSTRUMENT MAKERS	146
Aerial geophysical surveys	146
The Chobham navigator	147
Vehicular Magnetometry	148
Vehicular Radiometry	150
Down-hole magnetometer	150
Portable TEM system	150
GEOSTATISTICS	151
RADON GAS AS AN EXPLORATION TOOL	152
LEAD ISOTOPES AS GUIDES TO ORE	155
OPTIMISING GEOCHEMICAL DATA MANAGEMENT	158
DECREPITOMETRY	158

GEOMETALLURGY	159
UNDERGROUND NUCLEAR EXPLOSIONS	160
BETTER DRILLS	161
THE SEMINARS	162
THE MANUAL OF OPERATING PROCEDURES	166
'GEOSCRIPT'	167
SUMMATION	169

APPENDICES

Appendix 1: Extract from the McKinsey Report	179
Appendix 2: The Austirex story.	182
Appendix 3: Sir John Proud	202
Appendix 4: Value of Geopeko discoveries	205
Appendix 5: Summary of exploration campaigns	207

INTRODUCTION

This is the story of a mineral explorer, Geopeko Limited, a subsidiary of Australian mining house Peko-Wallsend Limited. Peko-Wallsend had grown from a struggling miner of a small copper mine in the Northern Territory into one of Australia's premier mining houses by virtue of a series of mineral discoveries made by its geologists in all parts of Australia. It had been formed originally from the merger of two small mining companies, one mining coal near Newcastle in New South Wales and the one mining copper near Tennant Creek, in the Northern Territory. The two merged in 1961 to form Peko-Wallsend Investments Ltd, and in the following year set up Geopeko Limited as the exploration arm of the group.

In the ensuing 20 years the group grew to an operation with mines in five states and the Northern Territory, metals businesses in Australia and overseas, and a range of complementary businesses. In the process the exploration teams made a substantial contribution to the understanding of ore formation and to the development of the technology needed for finding mineral resources.

This record of success had remained virtually unrecognised by the market until, in 1974, McKinsey & Company, on behalf of Conzinc Riotinto of Australia Limited ('CRA'), undertook a survey of management approaches to mineral exploration in Australia in the period 1950 to 1975. Thirteen companies agreed to participate. All had found orebodies in the study period and they were a mix of Australian-owned and overseas-owned companies, of large and small companies, of a range of annual spending on exploration, and with more successful and less successful records of ore discovery.

The companies provided data on the understanding that they would receive a copy of the report, in which the contributors were not at that time identified. Forty years on the data are now available and the companies involved have been identified. Geopeko is Company A in the figure below[2].

The results showed that Geopeko ranked highest in the gross metal value per exploration dollar spent and was amongst the lowest in the cost per find in exploration dollars. But perhaps the most interesting aspect of the results is the extraordinarily low level of recognition of the company's success by its peers (3rd column). Since PWL was a public company, and was bound to release promptly the news of its discoveries, it is difficult to understand how a company so successful fell below the radar in what is a very competitive field.

The McKinsey report went on to discuss the conclusions that arose from its study, which included an analysis of the reasons why some companies were more successful than others (Appendix 1).

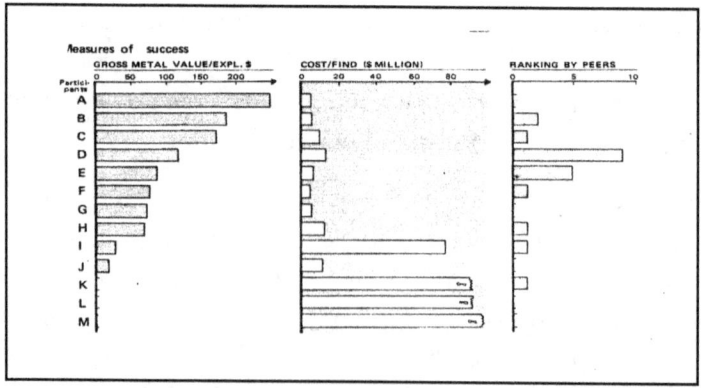

---

[2] Thanks to Riotinto for permission to use this diagram.

Geopeko was formed in 1962 but the foundations of the Peko-Wallsend group's success can be traced back to the years prior to the second World War when the Aerial Geological and Geophysical Survey of Northern Australia ('AGGSNA'), a Commonwealth body, carried out exploration on the Tennant Creek field as part of a country-wide campaign to improve the geological knowledge of the remoter parts of the continent. Tennant Creek orebodies are almost all hosted in bodies of magnetite, the magnetic oxide of iron, which throw a significant magnetic anomaly that is detectable at the surface by magnetometers, either on the ground or from the air. The majority of the ironstone bodies on the field do not reach the surface so magnetometry is an essential tool in exploration there.

A member of the AGGSNA team was geophysicist Lew Richardson, whose conviction that magnetometry would unlock the riches of the field, whose skills in developing the methods by which to do it, and whose conviction that the then fledgling discipline of geophysics should form an integral part of an exploration team, was a significant contributor to Geopeko's success, not just in facilitating discovery in Tennant Creek, and later elsewhere, but also in setting a high standard of technical performance in exploration around Australia.

After the War Richardson set up his own geophysical consultancy providing services to the Australian mining industry. He also persuaded business associates in Sydney, particularly John Proud, a mining engineer who he had known from their university days, to invest in exploration and mining at Tennant Creek, and Peko (Tennant Creek) Gold Mines No Liability came into existence.

Proud was convinced that the way to grow a mining company

was to find and develop your own orebodies, not buy them, and he believed that the actual exploration was best left to the earth scientists.

Peko then employed geologist John Elliston who, to that point, had been providing geological services to the company via a consultancy in Melbourne. He set standards of geological investigation that complemented Richardson's work.

To those three can be attributed the greater part of the company's subsequent success as they created a culture of innovation and scientific investigation that was embraced by the employees, both scientific and technical. Included in this culture was a drive to develop new geological concepts and exploration techniques.

Sir John Proud

John Elliston

Lew Richardson

CORPORATE HISTORY OF PEKO-WALLSEND

The town of Tennant Creek lies almost in the centre of Australia, in an area of low-lying hills made up of ancient sediments, now turned into rock, separated by wide plains where nothing crops out. Beneath those plains lies a host of bodies of magnetic ironstone which host deposits of gold, copper, bismuth, cobalt, silver, and other minor elements. Although gold had been found in one of the creeks that drain the field before the arrival of the 20th Century, the source of the gold was not identified until 1929. There followed Australia's last major gold rush as prospectors sought to find and test those ironstone bodies that stuck out of the ground.

Small mines were developed on those that carried enough gold, and one of those was the Peko orebody. In 1933 prospectors Joe Kaczinski and Bill Bohning had found an ironstone body protruding from the plains amongst the scrub ten kilometres east of Tennant Creek, applied for and were granted Gold Mining Lease No 102E, and named it Peko. Legend has it that the name came from their dog, which was using a Pekoe Tea chest as a home. If true, somewhere along the way the 'e' was lost.

At about the same time the Tennant Creek Development Coy. Ltd. was formed to test and develop mining leases on the field, and it acquired the Peko orebody. It developed it so successfully that in 1949 Peko (Tennant Creek) Gold Mines No Liability was floated in Sydney "to purchase from Mr J. S. Higgins the option held by him over the Peko Lease from The Tennant Creek Development Coy.

Ltd. together with two leases which he holds at each end of and adjoining the Peko Lease".

The gross value of gold proven to that date was £170,000. The float was successful and Jack Higgins stayed on as Mine Manager to see the new operation safely on its way.

Chairman was Sydney Sangster and the founding members were:

George Beattie Lean	Chartered Accountant	1 Share
Gordon Douglas Wharton	Stock and Share Broker	1 Share
Lorraine Marie Irene Pausey	Secretary	1 Share
John William Hansen	Clerk	1 Share
William John Baggie	Chartered Accountant	1 Share
Jill Clare Rogers	Stenographer	1 Share
Kathleen Tauchert	Stenographer	1 Share

Those who still held their shares twenty years later would have been members of a company with shareholder's equity of $204 million and total assets of $340 million: and which had grown, by discovery, takeover, and merger, to become one of the largest Australian-owned mining groups. By then it had mining and industrial operations in all Australian states and in several overseas countries, the value of annual production was $212 million[3], and it was the world's principal producer of bismuth.

Shortly after the acquisition the government, in order to assist the fledgling company and promote mining in the Territory, funded a drill hole to test a target deep below the Peko outcrop that Richardson's analysis of the anomaly had indicated was there. To everyone's surprise the hole encountered very rich copper ore as well as gold, thereby changing the history of the Tennant Creek field completely. At 320 ft the hole intersected approximately

---

[3] Peko-Wallsend Annual Report, 1978-1979.

10ft of lode assaying 14.6 dwt/ton (22.7 g/tonne) gold and 8% copper.

John Proud was also a director of the coal miner The Wallsend Holding and Investment Co., based in Newcastle in NSW. In 1961 he arranged a merger of the two companies, and Peko-Wallsend Investments Limited (PWL) was born. Proud had promoted the merger as a means of providing the group with a guaranteed long-term source of funding from the coal income that would underwrite the mineral exploration. 'Investments' was later dropped from the name.

To begin with, exploration was run from the Peko mine geology office but within a few years the company had discovered more orebodies on the Tennant Creek field and had also expanded into new geological provinces. It became evident that a separate entity was needed to manage it, so in 1962 Geopeko Limited was set up with its headquarters in Tennant Creek township. Later, as the company grew and the need for geological services to the growing number of mines increased, a separate division, the Mine Services Group, was set up.

By the mid sixties the teams had found four new mineable copper and gold orebodies around the Tennant Creek field: Orlando, Ivanhoe, Juno, and Warrego. These finds gave the Board the confidence to look beyond the confines of the Tennant Creek mineral field, not only to explore but to acquire other companies whose businesses complemented those of Peko-Wallsend. Among the first to be acquired was A F Toll Transport Ltd in 1961. In 1965 Tin Mines of Australia Pty Ltd and a half share in sand miner Rutile and Zircon Mines (Newcastle) Limited were acquired.

Geopeko's first exploration base away from Tennant Creek was established in Darwin in 1966 with the objective of exploring in the Top End of the Northern Territory. In 1967 the Australian

Government put up for tender a base-metal deposit, Woodcutters, near Darwin, that its geologists had found while exploring around the Rum Jungle province for uranium. PWL joined with Tasmanian lead-zinc miner EZ Industries to tender. The tender was successful and they formed the Gondwana Joint Venture to carry out exploration in northern Australia. The JV also took an interest in the South Alligator River Project, where a consortium of companies was hoping to resuscitate the uranium mining on that field where United Uranium NL, which had been mining there for over almost two decades, had fallen on hard times.

Also in 1967 two Australian mining companies, copper-gold producer Mt Morgan Limited in Queensland, and tungsten producer King Island Scheelite (1947) Ltd in Tasmania, found themselves in financial straits.

PWL saw off some overseas companies' bids for Mt Morgan and took it over.

But it found itself in competition with South African mining giant Goldfields in a contest for King Island Scheelite. Lew Richardson, following a visit to the operation with Proud, advised that in his opinion more orebodies could be found with geophysics, on the basis of which PWL raised its bid, Goldfields withdrew, and PWL took over the company.

Geopeko now found itself with two new geological provinces to explore, and two new mines to service, and bases were set up at Mt Morgan and King Island. In due course other provinces were chosen for exploration and new bases were set up in or near them from which exploration could be run. Bases were added, and later closed as the company's fortunes changed, in Parkes and Cobar in NSW, in Devonport in Tasmania, in Townsville, Brisbane, and Mt Isa in Queensland, in Perth and Kalgoorlie in Western Australia, in Ballarat and Ararat in Victoria, and in Adelaide in South Australia.

Operations were also set up in Chile, the USA, and other countries in later years.

As the company's activities expanded, however, it became clear that a national headquarters was needed so in 1968 Elliston moved to Sydney and set up a Geopeko office in the PWL offices in Lend Lease House, in downtown Sydney. But geologists and their rocks don't fit well into corporate surroundings so in 1972 the company moved its headquarters to Chatswood in north Sydney. Land was acquired in Gordon, also in north Sydney, headquarters were built to accommodate all of the company's headquarter activities, including the developing specialist divisions of geophysics, geochemistry, and computing, and the company moved there in 1975.

The group's growing financial strength also allowed it to continue to acquire other operations that complemented its mining interests. In 1969 it acquired the engineering firm Warman Equipment (International) Limited, with a variety of subsidiaries, and Gretley Collieries Ltd and H H Hardwick Transport Limited, all based in Newcastle.

The group's early expansion had been fuelled to a great extent by the geophysical expertise brought to it by Lew Richardson, his son Bob, and the team of geophysicists, mathematicians, instrument makers, and other experts that they had put together and who were consulting to the industry as L A Richardson & Associates ('LAR'). Later, following Lew's death, the group was sold to Peko-Wallsend, where it became the geophysical arm of Geopeko.

The discovery of the Ranger series of uranium deposits in the Northern Territory led to a contact with geochemist Geoff Sherrington, who had a uranium-focussed laboratory in Brisbane, and he was persuaded to join Geopeko to set up a geochemistry group in 1972.

At much the same time Geopeko geologists had begun to investigate the new discipline of geostatistics as a means of providing better resource estimates to PWL, particularly where very high grades of gold in Tennant Creek mines were likely to be encountered. Following consultation with the South African team who had developed the discipline Terry Quinlan was employed to set up and run a new resource estimation division that, of necessity, became the computer division.

These specialist teams later morphed into the Geoscience Division.

The discovery of Ranger One, a major uranium resource, prompted the formation, in 1971, of Ranger Uranium Mines Pty Ltd ('RUM'), jointly held by PWL and EZ through the Gondwana Joint Venture. In 1975 the Australian Government acquired a 50% interest in RUM, but when the Coalition Government came to power in 1975 the interest was sold back to the joint venture.

The discovery of uranium also prompted the Australian Government to ask Peko-Wallsend and Western Mining Corporation (WMC) to bid for contracts to explore for uranium in Iran, as had been requested by the Shah's government. After a series of meetings WMC withdrew, PWL bid alone and was successful, and in 1976 Geopeko subsidiary Austirex Aerial Surveys Pty Ltd was set up to manage the project. When the Iranian contract was terminated by the revolution Austirex was developed into a commercial aerial geophysical contractor and was later sold.[4]

On the Tennant Creek field the tight control of drill holes was essential, due to very difficult drilling conditions and the need to intersect targets whose position at depth was tightly prescribed by geophysical analysis. To begin with contractors were used but in

---

[4] Appendix 2.

due course the company developed its own drilling group, under the supervision of geologist Peter Kitto. As the group grew to service other bases Ron Kitching, who had built, with Jack Glindeman, one of Australia's premier drilling contractors and sold it, was persuaded to join PWL to run the group. It developed into a stand-alone drilling division, Overland Drilling, which was later sold. Also in 1978 the group acquired the metal trader Sims Consolidated Limited.

In 1978 the company set up the Mine Geology Services Group. Initially it concentrated on the Tennant Creek copper resources (to put on firmer grounds the decision to open a copper smelter in Tennant Creek)[5] but it was soon expanded to serve all of the group's mines. As part of a structural change in 1980 the Group and the associated Computer Division separated from Geopeko and were incorporated into the Metalliferous Mining Division ('MMD') of PWL.

John Proud had stepped down as CEO in 1974 and, now knighted, retired as Chairman in 1978. Following his retirement there was an important structural change in Geopeko. Centralised management was favoured as a way of getting more efficient application of resources and more rapid decision making. The bases were retained and remained under the control of Supervising Geologists who had much the same autonomy of spending on their existing projects as before but overall direction of the company's activities was vested more and more in head office.

There was also a period of PWL Board turmoil that later impacted heavily on Geopeko's operations. This was compounded

---

[5] Peko copper concentrates ran at around 25% Cu which meant that the company was paying for the transport, by road to Alice Springs and thence by rail to Port Augusta, of millions of tonnes of sulphide gangue. The logic of leaving that behind in Tennant Creek was inescapable. In addition, some Tennant Creek concentrates contained bismuth, cobalt, and other elements, and Peko was being penalised for it.

by weak commodity markets and the financial problems that emerged as the group's attempt to commission the flash smelter[6] in Tennant Creek failed. Elliston was promoted to a more senior position with responsibilities in other areas as well as exploration and Bob Richardson took over as executive in charge of exploration. But he departed in 1981, leaving Brian Williams to run the group's exploration until he was summarily retrenched in 1994. By then Peko-Wallsend had been taken over by North Broken Hill Limited.

Elliston left PWL in 1983 following disagreement about how it was being run and was promptly recruited by CRA as their exploration consultant, thanks to the McKinsey Report.

When PWL was taken over by North it was merged with North's gold company Norgold.

In 1986, with the acquisition of the Peak Hill gold mine in Western Australia, the group elected to set up a 'gold only' entity for taxation purposes and Peko Gold Limited came into existence. Past and present Geopeko geologists were recruited to run it. Services were provided to Peak Hill and the Kanowna projects near Kalgoorlie, and, after the merger with North, to the Bottle Creek Mine, also in WA, and to Geopeko discoveries in NSW.

Through the nineteen eighties financial problems caused Geopeko's funds to be severely cut and it was told by the parent Board to seek joint ventures or farm-outs or to sell projects. As a consequence it found itself working with partners on most of its operations, and in many cases diluting its equity. Thanks to its record in exploration it was able to remain as manager in most cases and keep the group together.

---

[6] There were two attempts to set up a flash smelting operation in Tennant Creek to reduce the cost of shipping product from central Australia, but both failed. A similar operation at Mt Morgan was successful.

It had, among others:

- Aquitaine farming into the Elliott Bay project in Tasmania;
- Anaconda farming into the Golden Dyke project in the NT;
- Renison Goldfields (RGC) farming into Mount Morgan in Queensland; and
- Chevron farming into the Lachlan and Goonumbla projects in NSW.

In some existing joint ventures, such as that with Shell in the Rover and Desertex projects peripheral to Tennant Creek, Geopeko elected to conserve funds by further diluting its equity. The Gondwana Joint Venture with EZ in the East Alligator Uranium Province embraced probably the most prospective ground in the province but the continuing uncertainty on tenure, due to the antipathy of the then Labour Government to uranium mining and the emerging Indigenous Land Rights movement, meant that even though feelers went out there was no sustained interest by other companies in farming-in.

There were also no takers for a Tennant Creek joint venture, even after the Board decided against in-house development of the Explorer 46 resource, which later became the Argo Mine, and it was added to the properties being offered. Here, and in most of the projects where Geopeko was in joint venture, field activities were cut back to preserve funds. There was also a concerted effort to sell those few surplus resources and tenements where potential was deemed too low to justify either a farm-out or continued exploration.

In late February 1983 then CEO Charles Copeman advised Geopeko of major a budget cut for 1984-85 year. The proposed PWL contribution of only $3 million demanded radical change in the scope of exploration and staffing levels. The reason for the cut was the need for $3 million to fund oil exploration. PWL had decided to acquire Weeks Petroleum Ltd, later to be renamed Peko Oil Ltd.

These cuts led to a severe reduction in the Geopeko establishment, with the loss of many experienced and competent personnel. A reduction of twenty two geoscientists, from a total of fifty, was proposed, seventeen by retrenchment and five by early retirement or transfer to mine positions. The announced changes brought morale to an all-time low. Plans to retain key experienced staff were undermined by the uncertainty and staff mistrust of PWL's management.

In July 1984 PWL held a strategy meeting to discuss the findings of an IBIS study on economic outlook and to develop strategy for the next few years. IBIS saw the likelihood of a severe downturn one to three years out and no real growth until about 1990. The PWL board view was even gloomier. Predicted return on investment was poor to unsatisfactory for most of its divisions, and several businesses, including Overland Drilling, were flagged for sale. Exploration spending, both mineral and oil, was to continue at its current reduced level. The corporate plan was to look to acquisitions to replace Tennant Creek within 3 years, but to support long-term commitment to exploration "so that on average there is a high probability that a new economic deposit will be discovered every 5-7 years". How that was to be done without adequate funding was not explained.

In 1982 PWL had acquired an interest in the Robe River iron mining operation in the Pilbara by buying Robe River Ltd's 35% interest in it. In January 1986 it took the opportunity to bring its equity in Robe up to 50.9% through the acquisition of Cleveland Cliffs' interest, and subsequently it took over management in order to improve project returns. There followed a prolonged but successful confrontation with the unions, who were supported by the Western Australian Labour government, court battles, and ultimately a change to work conditions, which quickly returned the project to profitability. Robe played an important role in PWL's financial recovery thereafter.

At about this time Geopeko reviewed its portfolio and its strategy on what metals and what orebody types should be targeted. It concluded that gold was top of the list with palaeoplacer, epithermal, and laterite-hosted the preferred models. Platinum ranked high, cobalt similarly. For copper, the only model seriously worth pursuing was considered to be stratiform copper. With an interest in two current tungsten exploration projects, Molyhil in the NT and Watershed in Queensland, and a worsening over-supply and price situation for tungsten, PWL policy on further exploration was seen as needed. For tin the only desirable target was Renison-type sulphide deposits. No expansion in lead and zinc was proposed beyond the current Jillawarra project in WA. Phosphate in Australia was considered worthy of review.

But the group needed a shorter time frame. Warrego was running out of gold ore and a new source was needed by 1986. Possible solutions were:

- an improvement in copper price so that Warrego and Gecko copper ore could be mined;
- development of shallow Tennant Creek oxidised gold resources to provide bridging until a larger deposit could be found;
- discovery by Geopeko of a mineable deposit elsewhere in Australia; or
- acquisition of an existing mining operation.

About this time a proposal was put forward by Geopeko that PWL approach EZ and North Broken Hill with the aim of merging the three companies into an entity of a size that could better compete with Western Mining, CRA, and BHP. It was rejected by the Board.

PWL was, however, looking to strengthening its position and was talking with Malaysia Mining Corporation (MMC), which had interests principally in tin, about possible exchanges of equity. Geopeko was included in a delegation of senior staff which visited Kuala Lumpur in March 1983 for a presentation on Peko's operations and plans to MMC senior management. The interchange between the two companies did not progress much further than that.

In 1987 North Broken Hill Ltd (NBH), which by this time had absorbed EZ and was again seeking to become a major mining house, made an all scrip offer for Peko Wallsend. Most senior Peko staff felt that Peko was the stronger company and that the takeover should be the other way round. For a while it looked as though there might be a counter offer but the Peko Board decided that submission was in the best interests of the shareholders, and by the third week in February 1988 the merger was complete, NBH would be renamed North Broken Hill Peko Ltd ('North'). By March the assessment of each other's resources and exploration portfolios was well underway but the two exploration groups continued to operate independently until North's subsidiary Norgold was absorbed into the parent company. All exploration was then, in 1991, consolidated under Geopeko.

Under the management of CEO Peter Wade and Mining Executive Ernie Miller North was supportive of Geopeko from the start. Geopeko had brought the Goonumbla (later Northparkes) copper-gold discovery to the advanced project stage and the Lake Cowal and Kanowna Belle gold discoveries were added soon after the merger. After the funding droughts of the early eighties exploration budgets reached the same level as in the late seventies.

Geopeko had continued to operate from Sydney after the merger, but early in 1991 the executive decided that the headquarters of both Geopeko and MMD should be relocated to North's head office building in St Kilda Road, in Melbourne. Since the merger such a consolidation had always been on the cards but the move had been resisted as much as possible. Quinlan continued in Sydney, giving support to the TORRES resource estimation software that he had developed, but with North's miners adopting commercial packages such as Surpac, and actively phasing out TORRES, he was retrenched the following October.

In April 1993 the Board initiated a skills audit of both mining and exploration. For Geopeko, the New York firm Anderson and Schwab Inc was awarded the contract. A team arrived from the USA and conducted interviews with a range of staff both in Head office and the bases. Questioning in Australia went in some strange directions. Williams had to go over the history of Geopeko a number of times. He wasn't questioned at all on technical understanding of the projects, on budgeting, on project review, or on systems in place to monitor progress, spending, training, research, and the like. Business manager Ian Calder was also mystified as to why he was not given the opportunity to explain Geopeko's staff assessment and remuneration system. Review at the bases seemed to have been more a technical assessment, as anticipated. The main feedback was the comment that there was too much use of computers.

The Anderson and Schwab audit report was received in Melbourne in late November. It was fairly damning on many aspects of Geopeko's approach to exploration, its culture, and its work ethic, less damning on its strategy and even complimentary on a few things such as technical ability of staff and husbandry of mining tenure. Early in the review Williams had expounded his view that exploration geologists work better when there is funding stability – when staff can see that the company is committed to exploration in the long term. Such a view, that stability correlated with better work, and the converse, that instability resulted in low morale and lack of focus, seemed alien to the auditors, who were very much in the then popular mode of "keep them perpetually worried about their tenure and they will work like beavers to keep their job". From this difference of view seems to have arisen the thesis that the auditors put forward, that Geopeko was a slow, not hard working, humanitarian organisation focused on "family", and interested only in the welfare of its staff.

What was even more concerning was that Geopeko management was criticised in the audit for using approaches, such as devolution of responsibility and accountability downwards, and computerisation looking towards better interpretation, that had been the very basis of its success.

In the wash-up there was one conclusion with which Geopeko had no problem - it had failed to sell the cause of exploration to the Board. Otherwise, why had there been such a wide-reaching review? It was equally clear that the Anderson and Schwab audit would be taken by the Board as a trigger for change in management. Geopeko's critique of the review was given to North in January but there was no time set aside for discussion of it, nor the recommendations from a management meeting that had been held at the time. North was restructuring at the top.

In February, changes were announced, including that Williams would take early retirement with Stephen Hare acting as Group Executive until a new appointment could be made. Williams was given a few days notice of his decision to retire. His replacement, Dr Aubery Pavard, ex Newmont in Denver, did not take up his position till November 1994.

A few months after Williams left, and after a small hiccup

when it was found that the name North Exploration was already registered in Queensland, Geopeko became North Exploration.

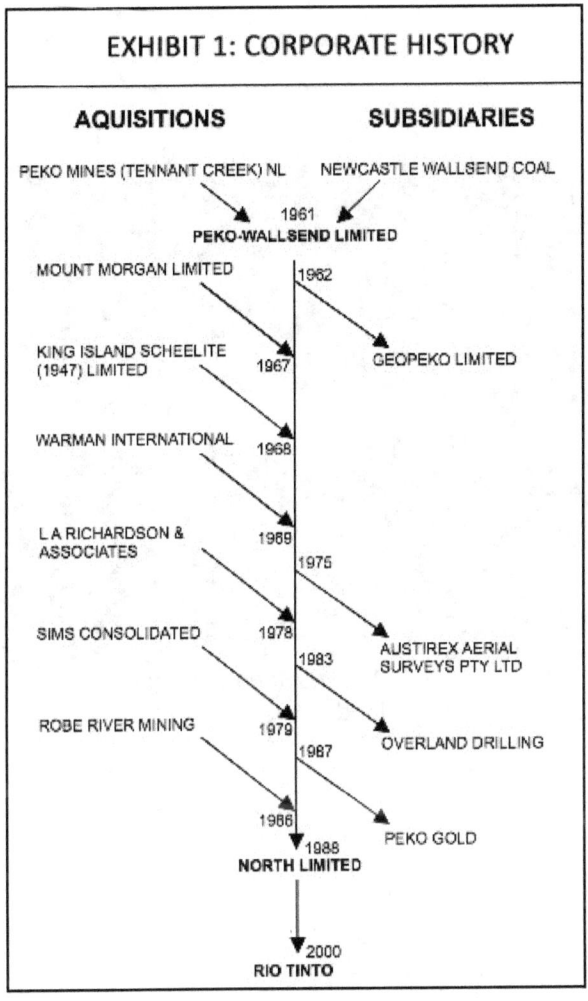

The hole that saved the company
Orlando, 1957

THE FOUNDATIONS OF PEKO AND GEOPEKO

Peko had begun life as a gold mine but in 1950 the drill hole that had been underwritten by the Bureau of Mineral Resources (BMR), which found the copper, changed the whole course of events. Until then metals other than gold had not figured in the field's potential.

John Elliston had been appointed as Peko's Chief Geologist, resident in Tennant Creek. His brief was to find more ore, with emphasis on copper and gold, and he embarked, in conjunction with Lew Richardson, on a campaign of identifying the most prospective magnetic anomalies on the Tennant Creek field, leasing them, and drilling them. Funds were very tight and some prospects, such as the Black Angel, which was later developed by another company, had to be let go. Despite the very slight understanding of the geological history of the Tennant Creek magnetite-hosted orebodies at that time it is an indication of their skills that the four mines that emerged from that campaign came from the first twenty anomalies that they selected, out of a total of the several hundred on the field that had been discovered by the aerial magnetometer surveys.

In 1956 Peko acquired leases over the small Orlando gold mine north-west of Tennant Creek and commissioned a drilling program on it. There had previously been some minor production – quoted as 46 tons of ore at a grade of 8.9g/t gold. There were two significant magnetic anomaly centres to the west of the old workings. It was not a strong anomaly but was of interest because it had surface indications of copper and was available cheaply.

It was the company's last gasp campaign, as there were no more funds for exploration. Had it failed it is probable that the name Peko would have disappeared from Australia's mining records within a few years. Fortunately it did not. The first hole encountered 36 feet (11m) at 26 dwt/ton (39.9 g/t) Au and the second 1.5 feet (0.46m) at 5 0z/ton (152.5 g/t) Au.

Much to the consternation of John Elliston, consternation that grew as early mine development failed to find mineable ore, the company elected to sink a shaft before there were sufficient drill intercepts of comparable value to be able to estimate a reserve with confidence and plan a mine. But in due course an economic deposit was outlined and in 1964 Orlando went into production, initially as a gold producer but later also as a copper producer.

In the meantime exploration continued elsewhere on the field and by 1964 mineable, or potentially mineable, deposits had been discovered at Ivanhoe, Warrego, and Juno.[7] All had economic copper grades but two, Juno and Warrego, also had very rich gold bodies with bismuth and silver credits, and it was those that laid the foundation for the company's expansion.

In 1958 the BMR had begun another program of mapping the Tennant Creek field, at a scale of 1:100,000, having earlier mapped it at 1:250,000. The program was concluded in 1959 and the maps became available in 1960. They provided a much clearer picture of the stratigraphy, and in particular the structure, of the sedimentary series that hosts the orebodies there. Furthermore, the original BMR mapping had raised doubts as to whether the porphyroids[8] that cropped out intermittently around the field were of igneous

---

[7] As Peko's first two mines already had names ending in 'o' it was considered an omen, and all subsequent discoveries at Tennant Creek were given names ending in 'o'.
[8] 'Porphyroid' – porphyry-like rocks.

origin, as had originally been thought, and this doubt was reinforced by the later mapping, and then by Peko geologists' research.

BMR geologist John Ivanac had also concluded that the porphyroids might have had something to do with the genesis of the mineralisation on the field. These doubts led Elliston on a forty-year campaign of research into the origin of these "igneous-looking" rocks, a campaign that led to a better understanding of the origin of ore-forming fluids, not only at Tennant Creek but generally, that undoubtedly contributed to the company's later successes.[9]

Elliston has published widely and continuously on his work. His hypotheses were radical to begin with and struggled to find acceptance in much of the geological community, but by and large were adopted by Geopeko geologists and certainly played a part in the group's success. But of more importance was Elliston's emphasis on looking at what any rock has to say about itself rather than blandly assuming that "because this rock looks like one I saw in Tasmania which is said to be igneous, then this one must be too."

One of the geologists in the 1958 BMR mapping party was Rob Ryan, who was working with the Northern Territory Mines Branch on secondment from the BMR. In the previous year he had been 'lent' to Peko to update the underground mapping at the mine. At the end of 1958 the Peko Board approved the employment of a second geologist, Ryan applied for the position and was employed, starting with Peko in early 1959. His principal role was as Peko mine geologist, releasing Elliston to oversee exploration.

The following year, as discoveries gave the company confidence in the future, a third geologist, Bob McNeil, was added to the

---

[9] Elliston, John, 2017. *The origin of rocks and mineral deposits*, Connor Court Publishing, Redland Bay, Queensland.

team as Project Geologist and took over responsibility for the daily management of exploration around the field, including the emerging discoveries of Ivanhoe, Warrego, and Juno, and in the Kimberley Division of Western Australia.

It was Elliston's policy that the company's geologists should be exposed to all aspects of its activities, so McNeil provided mine services to Orlando, Ivanhoe, and other discoveries, and Ryan from time to time was involved in exploration. This policy was expanded, as the group grew, to ensure that staff were rotated from base to base and from mine to exploration or vice versa, ensuring not only that they were well grounded in the company's activities but that they were exposed to all aspects of exploration, both geological and technical, to the practises of the mining industry, and the many types of mineral deposit that occur.[10]

Ryan left the company in 1962 for the Western Australia Geological Survey and was replaced by Kim Wright, who assumed the role of mine geologist at Peko. Brian Williams joined the company in 1964, and in 1965, when McNeil left, took charge of the group's exploration as it expanded around the NT and into the north-west of Queensland and northern Western Australia.

All the outside exploration activity was, however, straining relations with Peko mine management, which had to give up some of its scarce profits to support the geologists "running around the field in Jeeps". Matters finally came to a head in 1962 and the geological team was hived off into the new corporate entity, Geopeko Limited, reporting to the Board of Peko-Wallsend, not to Peko mine management. An office and accommodation were built in Tennant Creek and the exploration team moved off the Peko mine lease.

---

[10] It was considered of importance that the exploration teams should get a 'feel' for what an orebody looked like so that they would focus on those that had the greater potential.

The exploration geologists continued to provide geological services to the mines until the Mine Services unit was set up as a separate entity in 1963. That unit reported to the parent board in Sydney, not to mine managements, but a very close working relationship was established with the managements and the mine services geological team became an essential and welcome part of mine planning, development, and operations.

Critical analysis of the magnetic anomaly at Peko, using the methods they had developed, led Richardson and his team to deduce that there were extensions to that deposit that lay outside the known central body. A drilling campaign located a series of these and prolonged the life of the mine by many years. Over the years the same techniques expanded the resources at all of the group's Tennant Creek mines.

By 1961, following a new understanding of the geological features around Peko, the team had come to the conclusion that the porphyroids were of sedimentary origin and had recrystallised into igneous-looking rocks by a process which they were still to understand, but which was the origin of the mineralising fluids that had deposited the gold, copper, etc. in the magnetite bodies. This view was fortified by a drill hole below the Peko mine and above the porphyroid that exhibited clear evidence of the passage upwards along cleavage planes of mineralising fluids. An analysis of the dips and strikes of the area around Peko led to the suspicion that a porphyroid body extended continuously from beneath Peko, in a major s-bend, to below Nobles Nob, the other producing mine on the field at the time. The interpretation was tested and proved with some quick and cheap cable-tool drilling on the flats between the two mines, which found the porphyroid where it was expected to be, although, as time went by the stratigraphic succession proved to be more complex.

Drill core from Explorer VIII (Juno) - 1964

Lying along this sweep of porphyroid was a magnetic anomaly. It had heretofore been of little interest, as it was relatively small and had been drilled by another company, National Lead of America with, apparently, a negative result as nothing more had been done. It now became a target, was bought by Elliston, from the company that had the lease, with the contents of the Petty Cash tin[11], and drilled between 1962 and 1964. Drilling was extremely difficult, hampered by the propensity of drill holes on the field to swing to a direction controlled by the cleavage and texture of the host rocks rather than where the geologists want them to go[12], but by 1964 the target body had been successfully intersected and had proven to be a gold bonanza. It was named Juno, produced 454,997 tonnes of ore at 59.14 g/t Au, 0.38% Cu, 0.58 % Bi, with silver and selenium credits. The profits that Juno engendered drove the growth of the group for the next decade.

---

[11] The only money that was available.
[12] The failure of the National Lead hole was probably due to this problem.

PROJECT GENERATION

Project generation had traditionally been an exploration base function and responsibility. Darwin, Parkes, Townsville, Perth, Devonport, and other bases had been established not around an existing mine but as locations from which the exploration potential of selected regions, carefully chosen on the grounds of their perceived potential, could be assessed, acquisition opportunities monitored, both with the local prospecting community and the administration, and new projects generated. Spare time could be devoted to reconnaissance, thanks to proximity to the operating theatre, an asset not available to city-based exploration teams.

Regular visits by senior management kept track of operations and expenditure.

But by the late seventies success was taking its toll. Discovery brings with it demands for concentration of people and funds on economic assessment of the advanced project and on exploration in the immediate vicinity. Although a budget continued to be allocated annually for regional assessment and conceptual studies, too often people at the bases, at least the experienced ones, were too busy to be able to do justice to project generation and could not be spared from advanced projects already under way. Furthermore, the project development budget often needed to be reallocated to cover an unforseen shortfall in the funding of existing or newly discovered deposits.

The result was increasingly well-explored project areas, enhanced chance of local success, and staff well experienced in the local environment, but little opportunity for the knowledge gained to be applied in other regions of similar environment, and less still for the application of different genetic models in developing the next set of exploration targets.

In addition it became clear that a watch should be kept on the progress of exploration in the various provinces and that there would come a point when the probability of discovering new major resources in any active province had fallen to a point where the focus should be directed to new provinces. This shortcoming was recognised and the 1978 Geopeko management restructure provided for the setting up of a centralised Project Generation Unit.[13]

Brian Williams was given the job of planning, setting up, and leading the unit and he was joined by Andy Browne. Planning and start-up began in the Perth office but the unit moved to its permanent home in Gordon early in 1979. It had responsibility for assessment of all regions of Australia outside the existing project areas and their natural extensions. It operated independently from exploration base operations, reporting to the Exploration Executive, but it needed to work closely with the bases, gaining their confidence so that it could draw on local specialist skills, involve senior staff, particularly the Supervising Geologists, in reviewing and criticising ideas, and ultimately in selling the new projects so that they were picked up by the exploration teams. One of the first actions was a call for ideas from all staff. They were ranked, and follow-up was prioritised. The unit also needed to work closely with the geophysics and geochemistry specialist groups, in drawing on their expertise for interpretation of regional data sets, in assessing the practicality of exploring for targeted deposit types, in designing

---

[13] In 1978 Ryan had proposed, as the result of such an evaluation, that exploration should cease at Tennant Creek and move to new provinces where Geopeko was active. It was a recommendation that the company found too hard to accept.

the appropriate programs, and in a general overview of the project generation work.

With the consolidation of operations, first in Sydney and later in Melbourne, project generation became increasingly a function of Head Office, driven by the assessment of market priorities and by the financial constraints on the group rather than by geological assessments of potential and opportunity.

The unit oversaw the completion of Australia-wide reviews of tungsten and tin potential begun a year or so earlier, reviewed Australia's ultramafic complexes and recommended the Kimberley province of WA for exploration, involved a number of the base geologists in a review of stratiform copper potential and, among other studies, capitalised on the growing understanding by Parkes based geologists of the nature of the Goonumbla porphyry copper-gold mineralization, allowing the company to select similar geological environments in the central part of the Lachlan province. Exploration of those environments was taken on by the Parkes base as the Lachlan Project, which was rewarded with the discovery of the Lake Cowal gold deposit some years later.

Another gold-focused review was assessment of epithermal gold potential. This soon directed interest to eastern Queensland. With Geopeko's Townsville base closed not long before and the Mt Morgan base concentrated on VMS base metal search, there was no existing avenue to follow up this and other emerging potential in the state. A decision was taken in 1982 for Andy Browne to leave the Unit and set up a small base in Brisbane principally to generate and explore for this newly defined type of gold deposit.

With metal prices falling and exploration funding becoming more restricted the unit diverted some of its effort towards assessing the future commodity outlook in order to focus on what might be rewarding mineralization in the short term. In part this review was prompted by Board criticism directed towards Geopeko for its continued exploration for tin, not recognising the coming tin cartel crash. Review aside, it was soon apparent that the only commodity with potential in the short term was gold. One of the last projects of the unit was a review of alluvial gold in eastern Australia in the hope of outlining remaining resources that might be exploited quickly in order to shore up the corporate cash flow. Nothing useful was found.

The 1984 exploration budget cut was so severe that any thought of generating new large-scale exploration projects outside the immediate sphere of the few remaining base projects could not be contemplated and the Project Generation Unit was disbanded. As exploration via the bases gradually moved from project survival to normal and then discovery over the latter half of the decade, the bases took on the project generation role again, allocating the skills, time, and budget necessary to developing new ideas.

In 1993, following merging of the Norgold exploration staff into Geopeko, a head office Project Generation unit was again established, headed up by Peter Legge. It undertook a few broader scale reviews, adding support to the resumption of base metal search in the Carpentaria Basin, the reopening of the Darwin Base, and opening of a Mt Isa base, co-ordinated project generation work between bases, and developed an index of information systems available to the bases. But its most important contribution followed the decision by North to extend exploration overseas. All staff were asked to make recommendations of deposit types and host countries. It fell to the Project Generation Unit to assess these contributions, research the potential and the risks, and produce the short list recommended to management.

MINE SERVICES

Prior to the formation of Geopeko, geological servicing of the mines had been shared around the staff, who all lived on the Peko mine lease. To begin with services consisted of mapping underground exposure, planning, supervising, and assessing any drilling that was required to firm up reserve estimates and support mine development, and working with mine management to define resources.

Several things came together in 1963 to change that structure. With the formation of Geopeko only one geologist had remained on site at Peko to provide mine services but that changed about twelve months later and additional staff were recruited to cover the increased requirements both at Orlando and the newly discovered Ivanhoe. There was also the need for a separate exploration office as the space available at the mine office was inadequate for the increasing exploration effort and a need for a more dedicated mine group, particularly with respect to resource estimates that could be relied on. Furthermore, the need for more sophisticated resource estimation procedures where high-grade gold ore was involved was recognised. This became critical with the discovery of Juno, which had grades of gold that exceeded anything that had previously been discovered.

It was also recognised that it was extremely important that any additional reserves should be quickly identified and developed at the operating mines, as the infrastructure was already in place to support development.

The modelling developed by Lew Richardson played a major role in the prolongation of the life of the various orebodies.

As mining proceeded the magnetite envelopes of the deposits became more precisely defined by drilling and mining and by the measurement of their magnetic susceptibility. The geophysicists were then able to calculate exactly the anomaly that would be caused by that envelope and then subtract that anomaly from the observed anomaly to derive any residual anomalies caused by peripheral bodies of magnetite.

A major underground drilling campaign developed steadily over the next two years, at Peko in particular, to explore fully the whole mineralised body at each mine, including the satellite orebodies that had been identified by the geophysicists. That had the effect of increasing the known ore shoots in Peko from three in 1963 to eight by 1968. Peko mine alone was drilling core at a rate of approximately 30,000 metres per year for both exploration and ore definition purposes. This rate of drilling was also extended to the smaller deposits at Orlando and Ivanhoe. The drilling effort placed considerable pressures on core processing and storage. It was decided to keep as much of the ore definition core as possible in storage for future mining and metallurgical study if required. That proved to have been a very valuable decision later when research began on ways to distinguish ore-bearing ironstones from barren ones by their heavy metal isotope content[14]. The photographing of the core prior to splitting or sampling was also begun at this time.

Not only did the mine geological staff have to plan and service the drilling effort, they were also required to map all mine development openings and to inspect and mark up ore boundaries in the mine stopes for the miners. Day-to-day mining control was done jointly with the miners, mining engineers, planning engineers, and from time to time the metallurgical staff. Grade control was a day-to-day concern for all staff.

---

[14] *Lead isotopes as a guide to ore*, Page 72.

The discovery of the Juno orebody in 19624 was to change many things. After that discovery, surface drilling of new prospects was continued only until there were sufficient reserves to commit to mine development. The long surface holes were expensive and time consuming due to necessary deviation control requirements and the depth of many of the targets. The main shaft at Juno was sunk, and level access obtained, by mid 1967 when drilling could begin from underground. A series of horizontal holes was completed to locate accurately the ore ahead of mine development, and also to test for unclosed mineralisation, in particular to the west where analysis of the magnetic anomaly suggested there was more mineralisation. These holes located the No. 2 orebody, with one early intersection about 100m long averaging approximately 3 oz Au/ton (with high grade bismuth values also). An extensive grid-pattern ore-definition drilling program was then begun from underground, which not only defined the total resource but also disclosed that the Tennant Creek orebodies were zoned both geologically and mineralogically. That was an extremely important piece of information for future exploration in the field.

Ore reserve estimation

The calculation of ore reserves by traditional methods in the early 1960s was time consuming, tedious, and subject to considerable uncertainty, particularly where high-grade gold shoots and the nugget effect were involved. There was a constant effort to estimate the grades as accurately as possible, but reconciliation always produced a difference between those estimates and the metal recovered.

In the mid 1960s there had been a push in South Africa to address some of these issues through the application of new statistical techniques, named as 'Geostatistics'. The work was of sufficient interest that in 1968 Lois Jones, the wife of exploration geologist Gary Jones, and a mathematician, was employed to look at the application of the techniques in Tennant Creek. She set out to determine the number and distribution of grade populations in the mineral bodies. She demonstrated that the different alteration lithologies had quite specific grade populations, should be handled separately for grade calculation purposes, and should only be combined by careful volume weighting to give block, stope, and overall tonnages and grades.

That looked like adding substantially to the manual workload of calculation of the ore reserves, but fortunately technology came to the rescue. The first rudimentary digital/mechanical computers had become available, which permitted the calculations in a reasonable time frame. This was all just in time – the Juno No 2 orebody results were about to be fully disclosed to the public and senior management in Sydney wanted to be assured that the spectacular results being reported really were true. Dr H Sichel (one of the developers of the new techniques) was brought from Johannesburg to Australia and sent to Tennant Creek to check out the work and to advise the company regarding the quality of the results. That he went back to Sydney impressed was pleasing to all on site in Tennant Creek. Terry Quinlan and David Johnson were then employed to lead Geopeko into those areas.

At about the same time there was a full realisation within the Tennant Creek operations that "ore reserves" really needed the additional input of metallurgists and mining engineers to help define what was ore. Locally there developed the concept of "ore resources" and "ore reserves." The JORC Code was yet to be developed. However the result of all this was that the prediction of ore grades to the mill on a monthly basis improved to within 5% of actual recoveries.

Mt Morgan

Following the acquisition of Mt Morgan Limited the geological staff there were incorporated into Geopeko. Mt Morgan orebody had only a few years of ore reserves remaining so the principal effort was directed at exploration, with only a small requirement for mine geological services. The ore reserves heretofore had been calculated by the mine survey group and it was decided that that arrangement should stand. However Mt Morgan geologists Dick Reynolds and Grieve Brown made significant contributions to the mine geology, looking at the ore distribution and searching for any possible pit extensions. To this end a large model of the body was set up which displayed all the drill holes as rods coloured according to rock type and mineralisation. That enabled full three-dimensional viewing by all staff as an aid to understanding the ore distribution and also the relationships between the ore and surrounding geology. The practise was introduced on other operations around the country and proved to be of major assistance in demonstrating to the mining teams what the orebodies looked like.

King Island

When King Island was acquired by PWL in 1969 the only production was from No. 1 Orebody, which was mined by open pit, although the Bold Head orebody, 3km north of the pit, had been identified by consulting geologist Paul Anthony before Geopeko appeared on the scene. Little detailed geological mapping had been carried out in the open pit or elsewhere, and there had been minimal geological input into the mine planning process. A Mine Services team was put together and Paul LeMessurier was despatched from Mt Morgan to lead it.

By 1972 the Bold Head and Dolphin orebodies had been

delineated and mining had begun on them. Dolphin orebody, down-dip from No 1 Orebody, had been identified by Geopeko soon after starting exploration. It lay beneath the waters of Bass Strait, providing the geological team with new challenges in mine planning and operation.

Further, the team discovered that there were flaws in the services being provided to the mine such that there were serious differences between what was supposed to be arriving at the mill and what actually was. Three different survey grids covered the mine. Consultant Pat Coleman was brought in to set up a single accurate grid and this was then used as the basis of systematic drilling to define the ore boundaries and grades. The discrepancies then disappeared.

Until the arrival of Geopeko core had been logged by eye with no resort to UV lamping. It had been assumed that mineralisation was confined to certain rock types. The accidental exposure of some drill core in a tray to UV light, however, revealed that other mineral assemblages carried tungsten mineralisation, so a program of 'lamping' of the entire deposit was begun with the resulting delineation of the Dolphin orebody. Mark-up took place after midnight on most nights, when the team headed off for an hour or so in the cold and wet of a Bass Strait winter. It was a relief when all operations went underground. Because it was literally below Bass Strait considerable emphasis was placed on mapping the structural characteristics of the rocks. Planning of the underground operations was rigorous, to ensure that the mine did not flood.

The logging and mapping was carried out with the advice of the CSIRO rock mechanics group in Melbourne. Due to the thickness of the orebody – up to 80 metres – the post and pillar method of mining was adopted, and classified tailings were emplaced as fill. Each face of each pillar was mapped and photographed on each level. The mapped face was matched with the photograph, producing a continuous section for each of the pillar faces. Some pillars also had extensometers installed through their centre to monitor rates of failure and to link the movements with mapped structures and other weaknesses caused during mining. The general geology of the deposits was presented on 10m floor plans and 20m sections.

The rigid pillar system in the Dolphin Mine placed restrictions upon the location of mine faces and in the relatively shallow dipping, gently rolling sectors of the deposit. That caused some grade control problems when working 3m lifts from the 10m plans, due to the difficulty of accurately predicting, to the nearest metre, the position of waste within the ore body. The more erratic alteration of the dolomite at Bold Head also caused difficulty in grade control in the initial stages of that underground operation.

A major effort to control the waste and ore shipment from the mines was the "Fortnightly Booklet". This was produced by the relevant mine geologist and was designed to fill the data gap between the 10m floor plans with an up to date review of the geology expected to be encountered in the actual mining. The system worked well and allowed good communication between the various disciplines.

Tomago

RZM (Rutile Zircon Mines) was a joint venture between Peko Wallsend and Pioneer Concrete subsidiary Kathleen Investments Pty Ltd, formed in 1964 to mine mineral sand deposits for rutile and zircon.

In mineral sands the mineralisation occurs as very elongate deposits (strandlines) in which grade continuity in one direction can be hundreds of metres while in the other two directions is restricted to a matter of a few metres at best. These strandline deposits can contain very high grade cores (up to 40% Heavy Mineral) with a cross sectional area of maybe 2m by 5m but extending some hundreds of metres along the orebody.

Up until 1984 all the ore reserves had been calculated manually by survey and mine planning personnel using sectional grades, mainly at a drill spacing of 50 foot on lines 200 foot apart: and with, at that time, a cut off of about 1% Heavy Mineral. It was recognised that the number of intercepts of high grade cores recorded on the sectional drilling would be relatively low, and hence the high grade cores would not appear to be continuous from section to section. The ore grades estimated for the individual mining blocks would therefore, in some areas, seriously underestimate the actual grade that could be expected during extraction. One option would have been to close up the drill spacing along the lines to such a close spacing that the high grade cores would be intercepted on each line, but it would have been very time consuming and very expensive.

In 1984 therefore all Tomago data were entered into the computer at Gordon and run through the Torres system using a search ellipse very elongated in the direction parallel to the long axis of the strand lines. That produced a series of computer generated sections, plans, and grade blocks. The high grade cores were very clearly displayed on them, and close-spaced test drilling on the predicted position confirmed their presence. That allowed for much more precise mine planning and better scheduling of production. In 1985 RZM employed geologists for both mine and exploration work. They operated under the RZM name but reported to Geopeko Gordon for technical matters. Some years later Grieve Brown transferred from King Island and after the retirement of Bob Kelly, the General Manager, he assumed control until mine closure due to exhaustion of reserves.

Ranger One

The discovery of Ranger One, which included several uranium orebodies along a zone of disruption and alteration, highlighted the need for a method of ore estimation that would include a correlation between the geological characteristics of the ore and its grade. The team led by Terry Quinlan had Ranger One as one of its first projects.

With the emphasis on identifying ore of different characteristics, close collaboration between the site geologists and the geostatisticians was needed, and one of the latter was based on site. As a test, in the early stages a manual estimate and a geostatistical estimate were produced separately and compared. Statistical analysis of laboratory returns was also undertaken, with some eye-opening results, particularly the difference in assay returns on the same samples from different laboratories.

This rigour proved to be particularly important when the official near-by Nabarlek resource estimates were publicly written down by three-quarters and the board of PWL demanded an assurance that that would not happen with the Ranger One No 1 Orebody. It was at the time the only one of the Ranger One bodies for which an estimate had been done and published. The site team worked practically around the clock, from rechecking every assay return to final estimate, to confirm its estimate, which happily it did. When the No 1 Orebody pit closed many years later the estimate was found to have been within a few percent of the final production.

When Ranger Uranium Mines took over the operation Geopeko continued to provide ore reserve estimation services to it for some time.

Peko Gold

Peko Gold was set up by North at the instigation of Dr Ernie Miller to service the joint venture gold operations in Western Australia. Although not part of the Mine Services Group it performed the same function in Western Australia.

The first project to be taken on was the Peak Hill Main Pit, run by Peak Hill Resources under a joint venture between Forsyth/ Grants Patch Mining NL and Peko Gold. The second operation at that point was the gold operation of the Golden Valley Joint Venture between Delta Gold NL and Peko Gold and run by QED Mining. Work done by Terry Ballinger before the joint venture was established led to the discovery of Kanowna Belle which was brought to production by the Peko Gold team following on from the QED operation. A fourth operation, Norgold's Bottle Creek Mine, was included after the merger with Norgold.

All the geological services to these mines were provided by current or ex-Geopeko geologists and draftsmen, led by Peter Kitto and Terry Ballinger. They included the introduction of computer-based data management systems developed by Ballinger, resource estimation, and mine modelling, working closely with various consultants who were developing new software for resource estimation and mine planning, such as Surpac and Whittle 4D. This involvement led to improved mine performance, but also to some conflict with existing management when their management practices were shown to be lagging in current technology. Rigorous geological analysis of the data on various orebodies, via the Ballinger data management system, also led to the identification of new nearby resources in some instances, including Kanowna Belle, Fiveways, and Baxters. The team also supplied services to the Mesa J project at Robe River.

OPERATIONS: 1963-2000

As the group's confidence and assets grew it expanded to other parts of the Northern Territory, to the Kimberley Province in WA and to north-west Queensland; and then, progressively, to eastern Queensland, Tasmania, New South Wales, Western Australia, South Australia, Victoria, and later to several other countries. Geopeko bases were set up at the operating mine, if one had been acquired, but otherwise in suitable towns in or close to the provinces that had been selected for attention (Appendix 5). The philosophy was that the base should be as close to the target province as possible while being sufficiently developed that staff could bring their families and settle there.

Target provinces were chosen initially by senior management, but once a base was established the geologist-in-charge was expected to generate appropriate targets within the province and to run the operations from the base.

With the departure of Proud the emphasis changed, with more centralised control from Sydney and more forays into new territory based on board direction rather than the preferences of the geological team. That included moves into exploration overseas, although none of those had proved to be very successful by the date of the Riotinto takeover.

NORTHERN TERRITORY

Tennant Creek

Exploration continued throughout the Tennant Creek field at a steady rate from the formation of Geopeko until well into the eighties. It is estimated that some 250 magnetic features were assessed on the field. As knowledge and understanding of the geological features grew, and as exploration techniques improved, the company returned to prospects that had already been examined, and in some cases drilled. They included Black Angel, White Devil, TC8, and Argo. Gold and copper mineralisation was found in several others but at grades insufficient to make them economic.

Exploration campaigns were mounted as time went by on the Rover Field to the south-west of Tennant Creek; in the Granites-Tanami region; in the Harts Range and Jervois Range in the southern NT; in north-west Queensland; and continued in the Kimberley, including the Cummins Range, until the Pickands-Mather decision led to the abandonment of that program.[15]

In 1966 and 1967 the BMR completed another series of low-level aerial geophysical surveys over the Tennant Creek field. The data gave a much more detailed picture of the distribution of various lithological units which, coupled with Geopeko mapping, led to a better understanding of the field geology. New prospects that came from these BMR surveys included Explorer 38, located adjacent to the road from Tennant Creek to the Peko mine, and its neighbour, Explorer 46. Explorer 38 looked the better anomaly and seemed to offer the best potential, based on the rocks intersected in the early drilling but Explorer 46, an anomaly almost hidden in the background, proved to be the better target, and it became

---

[15] The Western Australian government granted exclusive exploration rights to Pickands Mather over large prospective areas of the Kimberley, thereby denying access to other explorers. This was at the time a new concept.

Argo, the last of the Peko mines in the field, over a decade later.

Through the period from the mid sixties to the early seventies Orlando, Ivanhoe, Juno, Warrego, and Gecko were brought into production.

ORLANDO

A production shaft was sunk in 1959/60 and a drive put out at the 380ft level (115m). It showed that mineralisation was not continuous (as had been assumed) and that each of the three holes had fortuitously intersected high grades, with the intervening ground containing little gold. This was a significant setback but further drilling from underground indicated that the level was opposite the top of a substantial and continuous body of mineralisation. Production commenced in September 1962 at 45,000t/annum. Orlando was developed as a satellite mine to Peko, using Peko processing facilities. In 1963 Ore Reserves (Resources as a term was not then used) were quoted as 250,000t at 11.7dwts/t (17.9g/t) Au.

Orlando had some of the most difficult underground mining conditions in the field, with a highly oxidised hangingwall to below 200 metres depth. Additionally the whole mineral body was quite large and it was felt that if an open pit mine could be established it would be an improvement. Consequently a program of drilling from the surface was undertaken to delineate the overall dimensions and character of the body. It located a considerable oxide copper resource but at the time (1965), with existing metal prices and metallurgical recovery technology, it was not possible to proceed with the open pit.

IVANHOE

Ivanhoe was a discrete magnetic anomaly pegged by Peko on a plain west of Tennant Creek, in January 1959, as Explorer IX. There is no outcrop nearby. Drilling commenced in November 1960. The first three holes intersected either low grade or narrow intersections but later drilling proved more successful and in 1965 Ore Reserves were quoted as between 250,000t and 370,000t at 4.5% copper and 3.0dwts/t gold (4.6g/t gold). Production began in early 1965.

Ivanhoe was also regarded as a satellite mine to Peko Mine and, together with Orlando, would probably not have been developed if the Peko processing plant had not been available.

WARREGO

The discrete and prominent magnetic anomaly that became Warrego Mine was acquired in November 1958 as Explorer V. There was no nearby outcrop. Hole 1 commenced in September 1959 but was eventually abandoned after it had strayed more than 60 degrees in azimuth from its planned target, even after the placement of six wedges. Hole 2 also deviated substantially in azimuth and was abandoned.

Lew Richardson on magnetometer survey with 'The Ferret' booking Explorer V (Warrego) - 1960

Hole 3 was drilled on a different azimuth, was collared in February 1962, and intersected 75ft (23m) at 3.92% copper and 1.95dwts/t gold (3g/t) between 990 (300m) and 1064ft (323m) down-hole; and 14ft (4.3m) at 3.22% copper and 0.8 dwts/t gold (1.2g/t) between 1096 (332m) and 1110ft (336.3m) down-hole. By 1965 an Indicated Ore Reserve of 1,370,000t at 3.2% copper and 1.7 dwts/t gold (2.6g/t) and Possible Ore at 200,000t at similar grades was published.

Although a decision to mine was made the following year, there were continuing concerns about its economics based on this resource. Further definition of resource to depth was impossible with diamond drilling because of the deviation problems so, in 1969, a light oil rig was brought in to drill vertical holes to just above the lode. Diamond drilling from these pre-collar holes confirmed the continuation of copper, found much improved gold grades and underpinned the economics. The mine was brought into production in 1973. Its copper resource prompted the two attempts at flash smelting at Tennant Creek. The second smelting failure in 1981 prompted a rapid escalation of mine development in order to define and mine the gold deep in the mine. This gold orebody made a major contribution to PWL's finances during the eighties.

JUNO

Following its acquisition the Juno magnetic anomaly had been modelled as a single ironstone body, although LAR geophysicist Greg Kater had considered that it was caused by two adjoining bodies. That interpretation proved correct. Luck intervened here as it was later found that if the original target had been intersected no significant gold would have been found and it is likely that Juno would have been abandoned as a relatively barren ironstone.

As it happened, after deviation the drill hole, in October 1964, intersected 40ft (12.1m) at 0.13% copper, 109.4dwts/t gold (167g/t), 1.39% bismuth and 19.7dwts/t silver (30.1g/t) between 752 ft (227.9m) and 792 ft (240m) down-hole − vertical depth about 680 ft (240m). A wedge runoff from Hole 2 intersected 51 ft (15.5m) at 187 dwts/t gold (286g/t).

Juno closed in 1976 having produced 838,942 ounces of gold, 2,275 kilograms of bismuth, 1,490 tonnes of copper, 89,767 ounces of silver, and 107,179 kilograms of selenium from 456,595 tonnes of ore.

GECKO

Designated Explorer I, this complex anomaly, lying 55 km north-west of Tennant Creek, was not drilled until 1967 although, as the number indicates, it was the first to be selected as an anomaly of significance. Detailed ground magnetometer surveys resolved the broad aeromagnetic feature into four separate anomalies, three of which had significant mineralisation. The discovery hole, at Anomaly 1, which became the Gecko mine, was drilled in October 1968 and was followed with the systematic drilling of 20 more holes in 1969 that proved up a significant new resource of 775,000 tons with an average grade of 3.65% copper and 0.4 g/t gold. Underground development began with the sinking of a cement-lined circular ventilation shaft in 1970. Drilling on Anomaly 3 in 1970 also intersected the first known significant primary magnetite-bornite mineralisation on the Tennant Creek field.

Production was interrupted twice by failures of the Tennant Creek smelter but over the ten year period until 1991 the mine produced 1.08 Mt at 3.8% Cu and 1.4 g/t Au.

WEST PEKO

Two kilometres to the west of Peko there was a broad magnetic anomaly that was obviously of interest. The target was deep. In it's early days Peko Mines could not afford to drill it so the NT government subsidised a hole on the basis that if it was successful the company would reimburse the government. In 1958 Glindeman & Kitching was contracted to drill a 1200-foot hole. It was slow going but in due course the hole was completed without intersecting any mineralisation. It did however intersect graded greywackes with well-formed magnetite crystals which were assumed to be the source of the magnetic feature, so the project was terminated.

With the advent of the down-hole magnetometer in the mid-eighties another hole was drilled, and logged with this instrument, in the hope that it would locate any magnetite lode that might have been missed by the original hole. A deep hole was drilled across the Peko Syncline, positioned to intersect what was thought to be the most likely position of any lode, and the magnetometer survey run indicated the presence of a magnetite mass. A follow-up hole intersected 25m of mineralised magnetite at a depth of 450m. The intersection contained copper at near ore grade but little gold. Step-out holes failed to find any better grades. Analysis of the data from this drilling indicated the presence of another magnetite mass to the east of the original discovery but this also proved to be of too low a grade and the project was abandoned.

BLACK ANGEL

The Black Angel anomaly was first drilled in 1957. There were encouraging results but the prospect was under option and Peko had no funds to exercise it, so it was dropped.

In 1968 Geopeko completed a 2-year option with Aurous

Development Ltd over an Authority to Prospect ('AP') that included Black Angel. Drilling resumed at Black Angel and White Devil without any further encouragement. Within the AP there were several magnetic anomalies of interest. They were given the series name 'Navigator' because Geopeko was reluctant to apply the Explorer name to a joint venture project in case it was inappropriately used for promotional publicity. The name was chosen because they were some of the first anomalies to be located on ground using the company's vehicle-based navigation system.[16]

Drilling of these anomalies proved to be extremely difficult due to the directional drilling problems that were common on the field. Navigator 1 swung through 94° of azimuth in 200 m! Drilling of the prospects intersected promising mineralisation in most cases, and it was known from earlier work that both Black Angel and White Devil were prospective. The option of $200,000 was due to expire so it was exercised and exploration continued, although through the ensuing years each prospect in the project area was eliminated as a commercial proposition.

The eighties gold boom saw a renewal of interest and the funding of a program of shallow drilling to locate readily exploitable resources throughout the Tennant Creek field. At White Devil a small resource of oxide ore was outlined from surface to a depth of 80m. Geopeko proposed that the resource be assessed for development and processing through the nearby Warrego plant, but PWL concluded that the resource was too small to justify modification of the Warrego circuits to take oxidised ore. Geopeko was advised to offer the project to Australian Development (by then a subsidiary of Poseidon) for processing through the Nobles Nob circuit, which was designed to treat oxide ore.

Tenders were called for further testing of the resource with the

---

[16] Page 68, *The Chobham Navigator*.

successful tenderer having the right to purchase the oxide resource and, if the seller, after review of results, was willing, to acquire the entire project. A local miner won the tender but he subsequently on-sold the right to Poseidon. The work was done and included a number of high-grade intercepts in oxide ore above the magnetite lode. It was concluded that they represented supergene enrichment, and Geopeko opted to sell the project.

As Williams notes "Some months later when Normandy was announcing the discovery of what became the million ounce White Devil orebody, we noticed that the position of one of the high grade intersections we had attributed to secondary enrichment had changed and, on their published map, was now in the top of the new orebody (a survey error, no doubt, corrected so as to not mislead the investing public)."

The nearest Geopeko hole had been 11m from the edge of the orebody.

Within only a few months of selling the Black Angel tenements the results of the Geopeko-sponsored CSIRO research work on the lead isotope signature of the Tennant Creek ironstones[17] became available. The signature of Black Angel, together with that of Juno, Peko, and Warrego, fell within the very small group ranked as having highest prospectivity. If it had been known at the time of the exercise of the option it would have been a strong incentive to retain and further test the magnetite lode.

Residual analysis on the Black Angel magnetic feature indicated that the White Devil lode plunged to the east, out of the Normandy tenements into ground still held by Geopeko. One drill hole was drilled near the tenement boundary to test the mineralization. It recorded ore grades in gold but at too great a depth to contemplate

---

[17] Page 72, *Lead isotopes and what they say about orebodies.*

developing a separate mine beside the White Devil operation. The tenement, holding the roots of the White Devil orebody, was also sold to Normandy.

WEST GIBBETT

The first hole at West Gibbet was drilled in 1959 and encountered good grade gold mineralisation with no evidence of copper. Since at that time Peko was focussed on expanding its copper resource the prospect was put on ice. Step-out holes during the sixties and seventies failed to extend the good gold values but they did outline a shallowly-dipping ironstone body, contrary to the steep dip modelled from magnetics. A deeper hole in the mid-eighties was aimed at what might have been rich gold deep in the ironstone, like that at Juno and Warrego. It intersected only minor ironstone and no gold. A down-hole magnetometer survey showed the ironstone to be much smaller than had been anticipated and that it had been fully tested by the existing drilling. There were no indications of concealed ironstones and the shape of surface magnetic anomaly appeared to have been influenced by an unusual level of remnant magnetism in the ironstone body.

ORLANDO EAST

Following the initial discovery and mining at Orlando, brownfield exploration resumed in the sixties. The first campaign located a shallow copper oxide resource within the Orlando Shear immediately to the east of the mine. The second focussed on the northern side of the structure and extended further to the east. It located ore grade gold mineralisation with pyrite and primary hematite, but no magnetite, in sheared sediments. A small resource was defined but PWL elected not to continue and the project was sold in 1991, as part of the sale of all Tennant Creek assets, to Normandy.

By the eighties the field had been well explored. Geopeko still had gaps to fill within the larger ironstones, seeking gold shoots that might be economic, but it needed a new crop of prospects. A more refined aeromagnetic survey flown by Austirex, and the newly available down-hole magnetometer, gave the opportunity to seek both hidden shallow ironstones with only subtle magnetic response, and deeper ironstones, their magnetic signature having previously been masked by the high magnetic background created by their host rocks. A number of both types were located and drilled. West Peko is an example of the second type. Discovery of the Orlando East gold ore associated with hematite, not magnetite, late in the eighties indicated potential for a third type of new target. But virtually all of the shear zones which would need to be explored were by this time held under exploration licence by competitors and there was not the corporate will to start again on acquiring a large holding in the Tennant Creek field.

In March 1990 there was the first indication that Normandy was interested in buying, or farming into, all of the group's Tennant Creek interests. From Geopeko's point of view the Tennant Creek Central Field was well explored and long delays could be expected due to Native Title applications. On the mining side, Warrego mine was at the end of its life and Gecko mine, with low gold credits, had always struggled to make a profit. Gold from the new Orlando East orebody offered a short respite, as did mining of the remnants from the old Eldorado mine, but there was no long-term resource to underpin continued operations. Re-treatment of the Warrego tailings appeared viable but this would be a separate operation not influencing the continuation of mining and milling.

There was no early follow up from Normandy and Geopeko

moved to bring in another joint venturer. By February 1991 a farm-in had been finalised with North Flinders and exploration was about to start when there was a formal offer from Normandy to buy all Tennant Creek assets for about $20 million. With North enthusiastic and Normandy urging, the sale was finalised in September.

Rover-Desertex

The host rocks to Tennant Creek mineralisation, the Warramunga Group, extend below Cambrian sediments well to the south-west of Tennant Creek, as indicated by aerial geophysical surveying. Geopeko extended its search into this area, naming the prospects 'Rover', in 1971, initially by farming into AP2451 which was held by Australian Ores and Minerals ('AOM').

This was a field where geophysics ruled absolutely. Richardsons undertook the original ground magnetic surveys to locate the first target, Rover 1, and closely supervised all field work, interpretation, and drill-hole targeting thereafter. Drilling discovered significant mineralisation at Rover 1 and eight other prospects, and was technically very successful given that the area is blanketed by 100-200 metres of Cambrian cover and all prospects were selected on the basis of the available aeromagnetic data.

Encouraging results from several Rover prospects, but mainly from Rover 1, over the next year or so prompted the company to take up exploration ground in its own right, firstly marginal to AOM but later extending further west, south, and south-east. Field work on the project, named the Desertex Project, began in 1973. In 1975, with exploration money tight, Shell was invited to farm in to Desertex, and did so. It later farmed into the Rover Joint Venture.

Shell's contribution underpinned further exploration around Rover 1 and initial drilling of some of the more remote magnetic anomalies, neither with much further success. Into the eighties copper price was again falling and exploration budget constraints restricted activities for both Shell and Geopeko. The joint ventures were finally wound up and most exploration tenements surrendered.

Metal prices began to improve towards the end of the eighties and the potential of the project, and Rover 1 in particular, again looked good. Exploration tenure was applied for and Geopeko entered negotiations with the Central Lands Council to reach agreement on access. They were unsuccessful and the applications and rights to negotiate were sold to Normandy in 1991. Normandy also failed to negotiate entry but exploration resumed in the first decade of the 21st Century when Westgold Resources and Adelaide Resources gained access.

Land rights impacts

Following gazettal of the Northern Territory Aboriginal Land Rights Act in 1976 the Tennant Creek Alyawarra Aboriginal Land Claim was lodged in October 1982. It covered much of the Tennant Creek central field and came right in to the town boundary. Peko decided to object to the claim on the basis that it threatened its continued mining and exploration activities; the NT Government also objected, principally on the grounds that a granted claim would prevent expansion of and supply of services to the town. The claim finally came up for hearing in 1984 in a modified form with much of the contested land removed.

Land claims were lodged over most of the Territory. This severely impacted on exploration around Tennant Creek. The Wiso claim covered all of the Rover and Desertex projects. Exploration was completely stalled for several years, negotiations with the CLC having failed to obtain even a minor dispensation to continue work on the partly tested prospects during the period of determination of the claim.

After much high level lobbying, in March 1982 Minister Tuxworth granted leases, under the new Mining Act, over Rover 1 and gave permission for the drilling of 4 holes at the Explorer 120 prospect. The land claim was granted in October 1983.

Towards the end of 1984 all Rover and Desertex tenements except for six Mineral Claims at Explorer 142 were surrendered and Shell and Geopeko were looking at terminating the JV.

Kimberley Province

Following a visit by Ryan early in 1962 to the Kimberley to inspect some tin and tantalite prospects that had been offered to the company by a prospector, Geopeko had acquired and begun exploration at Mt Angelo, a copper prospect near Halls Creek that Ryan was shown by the local station owner during his visit. Exploration in 1962 and 1963 resulted in the definition of approximately 400,000t of primary and oxidised copper mineralisation averaging 2.3% copper. Drilling was also completed at Ilmar's Prospect (so named after BMR geologist Ilmar (Bill) Gemuts who had been a member of the BMR mapping party in the Kimberley), and at Moola Boola, but no substantial mineralisation was intersected.

Exploration was under way when the Western Australian government granted to Canadian firm Pickands Mather exploration concessions over a substantial area of the Kimberley. Access to the areas of interest to Geopeko was severely restricted as a consequence, and it withdrew.

Tanami-Granites project

In December 1964 there was a brief geological expedition to The Granites by Elliston, Lew Richardson, and Williams. Richardson had participated in an AGGSNA geophysical survey there in the late thirties. There were several intense magnetic anomalies adjacent to the lines of old gold workings, and there was the possibility that, like the Tennant Creek field, there might be gold mineralization associated with these anomalies. Also, Geopeko held copies of plans and reports of pre-WWII exploration by a subsidiary of ASARCO, the US company with an interest in Mount Isa Mines during its early days.

The visit revealed remnants of the AGGSNA grid (a few wooden pegs not eaten by the termites and eroded to the shape of a knife blade by windblown sand), which meant that the grid could be reconstructed. Based on the knowledge of the geological characteristics of the rocks, it was decided to re-establish the grid and to drill one of the magnetic anomalies, Twin Hills, in the following season.

The old Tanami goldfield, 100 km to the north-west, was also a target for exploration but it had been held under mining lease by two Sydney investors for many years without any on-ground activity or formal exemption from labour conditions. Under the NT Mining Ordinance the holder of an inactive mining lease could apply to the Wardens Court for exemption, stating the reasons, and, hopefully, have it granted for one or two years. Otherwise, the leases could be over-pegged and claimed by another party ('jumped' was the term). Geopeko had no intention of jumping, as a previous episode at Tennant Creek had so alienated the local prospecting fraternity, with the consequent loss of the very worthwhile cooperation, that it was company policy not to jump. But it was felt that if some pressure was applied the Mines Branch might threaten forfeiture of the leases and the holders might not hold out against a sensible offer. Williams drove to Alice Springs, obtained formal searches on the status of the claims at the Registrar's Office, and Geopeko waited. The lessees applied for and were granted further exemptions and the company never did get to have a look at Tanami.

The drilling at The Granites recorded disseminated magnetite in schists, and assays of this material and the few conformable quartz zones, which had been auriferous in the adjacent small mines, showed no significant gold values. In 1967 little was done but the following year the decision was made to investigate the area more widely. It was a decision based almost entirely on geophysics. The published aeromagnetic maps showed a large number of anomalies ranging from discrete dipole anomalies, much like those at Tennant Creek, to linear and large complex features that looked like responses over mafic intrusives, and so proved to be. The geology was little known and, with outcrop minimal amongst vast areas of either sand plain or laterite-soil covered rises, there was little opportunity for the geologist to gain a better geological picture or to assess prospectivity.

During the 1968 field season a number of the anomalies were located and vehicle-born magnetometer surveys carried out. 'Pioneer' was to be the prospect series name. Pioneers up to about number 12 were located and, with no outcrop to help explain the source of the anomalies, a Gemco power auger drill was taken out late in the season and a few lines of widely spaced holes drilled to bedrock at three anomalies along strike from the Tanami workings. This work again produced no real indication of mineralization.

The 1969 season, and the next, saw the prospects reach Pioneer 35. Auger drilling was used not only for bedrock recognition but for grid pattern geochemical sampling. Pioneer 7, one of the anomalies found to lie over mafic-ultramafic rocks, received a lot of attention following auger drilling that revealed elevated nickel and copper values near the margin of the intrusive and the magnetic central zone. It was the time of the nickel boom in Western Australia and for a while Geopeko thought it might open a new front in the Centre. But it wasn't to be as the geochemical enrichment turned out to be just that, nothing more, and probably the result of lateritic concentration of the metals. Other drill holes were drilled at The Granites, at Pioneer 23, beside the old Ivy workings. There was one spectacular gold assay but insufficient encouragement for follow up. A distinct linear copper anomaly over Pioneer 17, in rocks similar to those at The Granites, was finally abandoned as too weak to indicate interesting mineralization.

Towards the end of the 1970 season there had been so little encouragement that another season's exploration was unlikely. The drills had been sent home, the camp battened down and all that was left was a little tidy up field work. Williams paid a visit to the Symington Ridge area 40km west of The Granites to look at some geological features that had been reported as interesting during some work in the previous season. He remembers being pleased to actually see some outcropping bedrocks. He sampled a few and they were assayed in Tennant Creek over the summer. One, and only one, recorded gold - about 6 dwt per ton (10g/t). By then, the project had been canned. There was some debate about taking a drill out next season to test this outcrop alone but, in that period when the Orlando mine needed ore to be 15 dwt/ton to be profitable, the significance of the Symington Ridge result was written down and there was no follow up.

In 1979 when gold price was finally rising, Geopeko should have been alert to the possibilities that The Granites and Tanami then offered but there had been a number of changes in staff and management in Tennant Creek and there was an inward focus which precluded looking for opportunities in the southern part of the NT. North Flinders Mining read the trend well and applied for exploration tenure over The Granites. Trevor Ireland and his team negotiated patiently with the CLC for years. They arranged a visit to Tennant Creek to look at Geopeko's drill core and asked about the sixties exploration. Finally they were successful in gaining access to the land. Williams gave them the pre-WWII data as a small boost to their initial exploration programme. The Granites mines were successful and, in 1987, discovery of the first of the Dead Bullock Soak deposits was announced. Dead Bullock Soak is 40km west of The Granites. That one sample not followed up in 1971 may just have started something much earlier.

Kurundi Goldfield

In 1964 Bill Cairns, owner of the Kurundi cattle station and part-time prospector, brought some quartz specimens into Geopeko's Tennant creek office. There was visible gold, and assays confirmed high gold values. Cairns had pegged a few leases and Elliston negotiated an option to purchase to get exploration under way. Assessment wasn't difficult, comprising geological mapping and outcrop sampling followed by a couple of short drill holes. The deposits were typical quartz veins with sporadic nuggets of gold, always difficult on which to make a confident estimate of gold grade. Geopeko's best estimate was about 5g/t, too low for mining at the then gold price. And the size potential was too low as well, so the project died within a few months.

Strangways Range

The early BMR low level aeromagnetic surveys which had sparked a new round of exploration at Tennant Creek in the sixties also aroused Geopeko's interest in the Strangways Range, in the Arunta geological province, north of Alice Springs. Two of the strong aeromagnetic anomalies were associated with known copper mineralization (Johnnies Reward and Kennedys) and there was a strong anomaly over the marble at Mud Tank, which had been recognised, by David Gellatly working with the BMR a few years earlier, as the first Australian example of carbonatite.

In 1965 Geopeko decided to investigate the possibility of a Tennant Creek analogue, copper-gold ore in a magnetite host, in these Arunta rocks. Agreements were negotiated with the lease holder of Johnnies Reward and Kennedys and a team located a number of aeromagnetic anomalies, including the Mud Tank anomaly, and read reconnaissance magnetics over them. Johnnies Reward was diamond drilled with both holes recording some copper and gold but not of economic grade. Several of the anomalies located by the reconnaissance surveys, including Redrock Bore and Harry Creek, were followed up with detailed magnetic surveys and considered as possible hosts to mineralization, but in the light of the disappointing results from Johnnies Reward didn't rank highly enough for a drill test.

Mud Tank was different. In Africa, mineral deposits associated with carbonatites were being mined for a range of commodities, the largest being the Phalabora copper deposit in the Limpopo Province of South Africa. With a large copper deposit as the potential goal, Geopeko set out to explore the Mud Tank intrusive. There were marble-like rocks containing crystals, large and small, of titanomagnetite, apatite, and zircon and, between outcrops, a surface lag layer of these three minerals well known to fossickers and mineral collectors. Micas were also common components of the carbonatite and their presence in the soil together with the shiny crystal faces of the lag gave the hill a bright sheen in the sunlight.

Exploration soon eliminated the possibility of a large copper deposit but, with little phosphate resources known in Australia at this time, the northern part of the intrusion was considered a possible source. Here, apatite was much more common, making up 30% of more of the surface carbonate rocks. Two diamond holes were drilled that showed the nature of the fresh carbonatite but also revealed conclusively that the apatite at surface was a gravity concentration of weathered-out crystals re-cemented by secondary carbonate. The true phosphate content of the carbonatite was perhaps three per cent, much too low to be of interest as a source for fertilizer. Similarly, zircon, which could have been a valuable by-product if there had been a viable apatite mine, made up well below one per cent of the primary rock and could not be considered valuable in its own right. The project drew to a close in 1967.

Fitzmaurice River

In the mid-sixties Tennant Creek Base was looking for opportunities for expansion into new exploration projects anywhere in the Northern Territory or in North West Queensland. A few years later, with the opening of the Darwin Base and the forming of the Gondwana JV with EZ Industries, a line separating the jurisdiction of the two groups was drawn at 18°S. Geological reconnaissance from a light plane was used in some areas where no 4 mile to the inch mapping existed, and it was on one such reconnaissance that Elliston and Williams saw outcrop of possible felsic porphyries near the Fitzmaurice River. It was decided in 1965 to apply for an Authority to Prospect between the Fitzmaurice and Victoria Rivers to check the area for possible similarities with Tennant Creek. Aerial photographs were obtained, base maps prepared via the slotted template method, and photo interpretation was undertaken.

The interpretation concluded that there were inferred porphyries, associated sedimentary rocks, and granite in a small erosional window in flat lying sedimentary rocks near the Fitzmaurice River. These flat lying rocks extended out well to the east. To the west in the Bradshaw Range there were unrelated steeper dipping sedimentary rocks and further west again another seemingly unrelated unit also probably sedimentary. Within the window there were a few prominent dark outcrops which could have been lodes so a field visit was planned. A small A to P was retained over the window of older rocks and later flown for aeromagnetics but this revealed no interesting anomalies and the project was terminated.

Oonagalabi Project, Harts Range

The Harts Range had long been known for its mica mines, all of which had ceased production by the sixties, and there had been a number of small occurrences of copper and minor metals as well around the Huckitta Dome. The discovery of extensive surface copper, lead, and zinc mineralization at Oonagalabi on the southern side of the ranges by geologists Ken Neilsen and Ian MacCulloch of Russgar Minerals in early 1971 came as a surprise and gave early promise of becoming an important ore deposit, a first for this province. Geopeko was invited to inspect the prospect after some months mapping and sampling, electromagnetic and magnetic geophysics, and the drilling of 14 percussion drill holes by Russgar. A farm-in was negotiated and a Geopeko team continued exploration of the prospect on a campaign basis from Tennant Creek during the winter of 1972. The mineralization was mainly stratabound zinc and copper in high grade metamorphic rocks. The possibility was that it represented a metamorphosed volcanogenic massive sulphide deposit, or perhaps a Broken Hill analogue. The Russgar exploration had downgraded the near surface potential for high grade mineralization and Geopeko concentrated on assessing the possibility of more substantial mineralization at depth.

It became apparent that the chances of there being a substantial deposit still to be found were small and the company withdrew from the venture. Amoco and others subsequently extended the exploration but ore grades were never sufficient to make mining the deposit viable.

Jervois Ranges and Molyhil tungsten

Geopeko's interest in rocks of the Arunta Province in the mid-sixties extended as far as the Jervois Ranges, 300km north-east of Alice Springs. The Jervois Copper mine was considered to have too small a potential to be of interest but there were a number of field trips seeking prospects that might meet the company's criteria. They were mainly prompted by prospectors bringing in samples from newly-found copper showings in the Bonya plains where they seemed to abound. It soon became apparent that this was a field of "squibs" and interest died.

It was renewed in 1981 when the Molyhil tungsten-molybdenum deposit was offered by Petrocarb and Nicron who were seeking a joint venturer to advance the exploration. Geopeko was then on a nationwide campaign to acquire good tungsten resources to complement the King Island operation. Farm-in terms were reached and work began in October. The Molyhil deposit consisted of scheelite, molybdenite, and some chalcopyrite and pyrites in a magnetite-calcsilicate skarn. Exploration of the deposit was well advanced and Geopeko's prospect work involved mainly better defining and firming up the resource. Considerable effort was directed at improving grade to enhance project economics. Metallurgical testing, mostly entailing ore-sorting, was undertaken off-site by Peko's Metalliferous Mining unit. Surrounding the prospect was a large block of exploration licences held by the joint venture. It was systematically explored, using firstly aeromagnetometry, then field location of the magnetic anomalies utilising 'Blossom'[18], followed by prospect level sampling.

During 1982, as a consequence of uncontrolled sales of tungsten ore out of China, wolfram and scheelite prices dropped radically. The King Island operation was suffering financially and in November one operating shift was dropped and the workforce reduced by 85. With no foreseeable improvement in tungsten price, the goal of making Molyhil economic receded into the distance. Field exploration continued through the summer with anomaly location and checking, but with no new mineralization indicated. The decision to withdraw from the joint venture followed soon after.[19]

DARWIN AND JABIRU

As the company's operations grew negotiations were also begun with other companies with a view to joint venturing exploration elsewhere in the Territory. As a result it was decided to open an office in Darwin. Coincidentally, Ryan had written to Elliston proposing that Geopeko begin exploration in the Pilbara region of WA, where he had been working and where he saw a similar potential in some Archaean sequences to that at Tennant Creek. He was offered the Darwin position instead, and accepted, arriving in Darwin in January 1966, where he was joined by Eric Swarbrick, also from the WA Geological Survey. In due course an office was set up in Winellie. Adopting the series name 'Quest', the team set

---

[18] See page 68 for an explanation of 'Blossom'.
[19] To induce Nicron to accept the withdrawal Geopeko offered Woodcutters (see this page) as a sop, and it was accepted.

out to identify and acquire base metal targets in the Pine Creek Province.

Exploration had barely begun out of Darwin, however, when two projects fell into the company's hands: Mount Bundey iron ore mine, as the result of the takeover of Mt Morgan Limited by PWL; and Woodcutters, a lead-zinc-silver discovery by the BMR at Rum Jungle during its exploration there for uranium, which it had put up for private tender.

Mt Bundey

The Mount Bundey Mine, a modest hematite deposit, was already in operation under the management of Alan McIntosh for Mount Morgan Limited, and Geopeko took over the provision of mine services. Production ceased in the early seventies and the mine team transferred to Jabiru to take over responsibility for the development of Ranger One, the uranium orebody that Geopeko had found.

Woodcutters

The Peko-EZ tender for Woodcutters was successful and the project came under the management of the Darwin office. With it came geologist Alex Taube, who had worked on the project with the BMR and wanted to continue. With new JV funds available a vacancy existed and Taube was brought on board. Over the next few years he ran the project and outlined a significant resource in what was a geologically difficult environment.

By 1981 the project had been advanced to the feasibility stage and found to be non-viable. Attempts to sell it had brought little interest and surrender of the leases was being considered, but interest from ex-Geopeko staff Bob Richardson, Ryan, and Ray

Soper, and Geopeko's need to escape from the Joint Venture with Nicron Resources, led to sale of the project to Nicron and Lachlan Resources, who brought it into production.

Alligator Rivers Uranium Province

Work had barely begun at Woodcutters, however when, in 1968, the BMR released an updated version of its Pine Creek map, which resulted in a re-evaluation of the mineral potential of the country between the South Alligator River and the East Alligator River, about 200 km east of Darwin. It suggested that the area, which had a few granitic outcrops that had been assigned originally to the Proterozoic as intrusives but had been dated as Archaean, was analagous to the Rum Jungle mining field where uranium and base metals had been mined, and was hence an exploration target. Authorities to Prospect were applied for over what was virtually nothing but buffalo pasture. A reconnaissance in the 1969 dry season showed that outcrop in the area was practically non-existent. It was apparent to Geopeko that exploration would have to rely heavily on geophysics, so the Joint Venture applied for a much larger area and wrote a contract with contractor GRD to fly a radiometric and magnetometric survey of it in 1970, once the monsoon had abated.

But in October of 1969 Canadian explorer Noranda contacted the Joint Venture with the news that GRD, while carrying out a survey for Noranda on its adjoining licence area, had detected radiation on Gondwana ground, had carried out some reconnaissance, and had identified a very strong radiometric anomaly. Coordinates were given. A Geopeko team located the discovery and pegged a mining lease over it, it being Geopeko policy at the time that no serious exploration should be done except on ground held under lease and it was clear from the strength of the anomaly that there was a major accumulation of uranium there. The series name 'Ranger' was adopted and it became Ranger One, which was in due course developed by Energy Resources of Australia into the Ranger Mine.[20]

Early in 1970 a ground team was flown in and exploration begun at Ranger One. It soon became clear that the Joint venture had a major uranium discovery on its hands and by the end of the year a permanent base had been established and named Jabiru, the aerial geophysical survey had been completed, and exploration was planned over the Joint Venture's licence areas under the designation of 'Munmalary', the name of a local buffalo station. A separate team was set up at Jabiru under Ron Lees to manage that program. Faced with almost no outcrop over the entire area, exploration was driven by geophysical mapping, soil geochemical sampling, and innovative drilling that allowed the low-cost recovery of core from the basement beneath up to 100m of cover. Several highly prospective targets were generated, not only for uranium but also for precious and base metals, with the help of sophisticated analysis of the geophysical and geochemical data.

By 1975, when Prime Minister Whitlam visited the Geopeko operation at Jabiru, it had become clear that the Whitlam government would not support uranium mining, and activity at Ranger One was winding down. Further, the proposals to create the World Heritage Kakadu Park were threatening the Munmalary program, and it had virtually ceased, despite the growing evidence that a major mineral province was being unearthed.

A change of government in 1975 brought some relief and Ranger Uranium Mines Pty Ltd took over the development of Ranger One.

---

[20] The action of pegging the lease may very well have guaranteed the future of the discovery, as many other targets outlined by subsequent work have been locked up in the Kakadu National Park.

The new Coalition government also indicated to the JV that if it applied for leases over its discoveries elsewhere in the province they would be excluded from the park, so in 1976 Geopeko carried out a major rush campaign aimed at identifying prospective ground and eliminating unprospective ground. Mining leases were applied for over what was thought to be the most prospective features around the field. The government undertaking was not honoured, however, and the Munmalary program was shut down at the end of the year.

Prospects that were identified included:

- Ranger 68 (Barote Springs) an ore grade deposit of uranium;
- Ranger 4, also potentially an economic uranium deposit;
- Ranger 3, where nickel showings were located in association with a mafic rock and a magnetic anomaly in the Archaean;
- Ranger 34, which had base metal mineralisation in association with altered dolomite; and
- Ranger 43, which had traces of gold.

Several other uranium-sourced radiometric anomalies were also identified in the area.

The Atomic Energy Commission, at the instigation of Rex Connor, the Labor Minister for Resources in Whitlam's government, had also carried out exploration in the area and located uranium mineralisation at Austatom 1.

Cooper Creek and Borrodaile

In 1980 the NT Government took the first steps towards renewing (or reallocating) the frozen exploration tenements; it arranged a large multi-sponsor airborne geophysical survey of the province and it opened up Arnhem Land for exploration under an Exploration Licence tender system. It was clearly stated that tenders would be considered on the basis of technical merit. Geopeko made the decision to participate with the aim not of gaining large holdings but of obtaining the very best ground. Based on what it knew from its Munmarlary Project work it selected two conjoined exploration licence-sized blocks in north-western Arnhemland, logistically difficult to explore because the lowlands are covered with water or saturated through the wet season and for some months beyond, but interpreted as containing extensions of the highly prospective rock units around Ranger 1, Jabiluka, and in the Barote Springs block. They were given the names Cooper Creek and Borrodaile.

The submission prompted a compliment and a statement from the NT Government that Geopeko's model would be used to solicit quality submissions from other applicants.

In July 1982 Geopeko was advised informally by the NT Government that it would be offered only a small part (about 12km^2) of the area applied for. The winning entity, a joint venture between Mobil Energy Minerals and Suttons Motors, had proposed spending $1.5 million in the first year of exploration and $10 million over 5 years. Geopeko's proposed expenditure came in at about the middle of 22 submissions. It considered that the proposed level of expenditure in the first year was ridiculous, as most would be wasted considering the very short field season. The area to be offered was much too small to warrant mobilisation of people and equipment so Geopeko withdrew.

In early 1983 Mobil approached PWL with a proposal for Geopeko to take a controlling interest in the Joint Venture. Attempts by Mobil-Suttons to gain access to the Aboriginal land had reached a stalemate and the Joint Venture was seeking Geopeko's assistance. It was agreed, the necessary documentation was completed, and

Geopeko joined Mobil's Lyle Hennage in a series of meetings with the Northern Land Council ('NLC'), Chief Minister Tuxworth, and the NT Mines Department trying to negotiate terms for entry, exploration activities, and mine development.

Negotiations dragged on through 1983 and most of 1984. Mobil was getting weary and threatening to withdraw. Geopeko similarly had lost much of its enthusiasm. The culmination came in November when Hennage and Williams had dinner with Galarrwuy Yunupingu and John Ah Kit of the NLC in Darwin. Lyle stated the facts. Both companies were suffering financially and losing experienced uranium explorers with the delays. Mobil had to have a deal very soon or it would walk away. Both of the guests burst out laughing and said that they didn't believe it. It was only a negotiating ploy. Two weeks later Mobil closed its office in Sydney and Hennage was on his way back to the US.

In March 1985 Mobil advised that it was shutting down Mobil Energy Minerals and was prepared to sell its tax losses and/or its property rights. In April Darwin office was told that the NLC was ready to talk about access to Cooper Creek but there was no functioning joint venture and the invitation could not be followed up. The following January Mobil announced that it was selling its rights in the Cooper Creek JV to PNC and Stuart Williams, PNC's exploration manager, met with Geopeko. The sale process dragged on and it was June 1987 before Peko and Suttons had signed the sale agreement. It proved difficult to find someone in Mobil able to complete the sale but the new joint venture did come into existence later in the year. Geopeko was to be the manager.

Exploration aimed at getting started in the 1988 dry season was planned and budgeted, consideration given to minimising environmental and social impacts, and a proposal prepared for submission to the NLC. The venturers were encouraged by the smooth progress through the NLC to an arranged "bush" meeting with the traditional owners set for April 11. The presentation was made, the venturers retired to find a shady tree and some time later were called back to be told that the proposal had been rejected. Under the Act, it would be 5 years before another proposal could be considered.

By 1993, Terry Hoschke was leading the small Darwin Base team exploring for gold and base metals within the Pine Creek Geosyncline. He flagged the need to reactivate the joint venture for another attempt at gaining access to Cooper Creek and Borrodaile. This time PNC offered to prepare the submission which, as before, included details of proposed exploration, minimising environmental disturbance and methods of rehabilitation, proposed compensation to the landowners, and minimising social impact. The submission was lodged in April, presented to the traditional owners a few months later, and again rejected.

South Alligator River Joint Venture

The South Alligator Joint Venture had been set up in the early seventies to bring in funds for continued uranium exploration by the struggling United Uranium NL, following depletion of the resources in its South Alligator mines. After a visit to the operation by Proud, Elliston, and Ryan Peko and EZ decided to take up an interest. By the nineteen eighties there were six co-venturers, Utah, Dampier (BHP), Noranda, Newmont, and the Gondwana Joint Venture, some contributing, some diluting, and a long history of no success.

It was time for rationalisation: reduce the number of venturers or abandon the joint venture. Dampier wanted to contribute no further funding; Utah wanted to dispose of its interest; Noranda was undecided but, in the interests of rationalisation, offered $50,000 for the Utah-Newmont-Peko-EZ 55%; Utah offered half of its 24% each to Noranda and Peko-EZ.

Nothing was finalised at the meeting but Noranda's surprising offer induced Geopeko to have a good look at the assets of the joint venture. It soon became obvious that there was gold in the roots of the old Coronation Hill mine at grades not interesting in the sixties but clearly interesting now. EZ was informed and both EZ and Geopeko set about getting $12,000 to take up part of Utah's interest. PW CEO Copeman wouldn't support the proposal and Peko withdrew from the joint venture. There was consternation in the mining community as the resource built to over a million ounces of gold and the consortium watched in dismay as the value dwindled to zero due to Indigenous opposition, the declaration of Kakadu National Park Stage III, World Heritage listing over the South Alligator Valley, and the Australian government's refusal to grant permission to mine to the project. The project did not proceed.

Burrundie

Following the loss of the Munmalary project, exploration resumed over the rest of the Pine Creek Province, assisted by the latest BMR mapping, with emphasis on volcanic-hosted massive sulphide deposits. A camp was set up near Burrundie Siding on the North Australian Railway with John Goulevitch in charge, and targets set out by the geophysicists were systematically explored. They had re-interpreted previous magnetic data and drilling results to demonstrate that the base-metal and precious metal prospects had not been satisfactorily tested by earlier explorers, and that there was significant potential for magnetic ore at depth in both cases. They recommended and targeted drill-holes that led to the discovery of the main mineral bodies of 980,000 tonnes at Iron Blow and 550,000 tonnes at Mt Bonnie, both deposits that had been mined before. The oxide sections of both mines and part of the sulphides at Iron Blow were mined in the mid-late 80s, but by other companies. By then Peko had disposed of its interest.

The Cosmo Howley gold mine was offered to Geopeko in 1977, but with gold still below $100/oz the offer was declined, to be taken up later by the Anaconda JV.

By the end of 1981 much of the programme had been joint ventured with Anaconda as the Golden Dyke JV. Focus was largely on Cosmo Howley. The Mt Bonney and Iron Blow deposits did not meet Peko's minimum mining size and were sold. An offer of the Woolwonga prospect was taken up by the Golden Dyke JV. Budget cuts in 1984 led to a total write down of Geopeko's activities in the Top End. It continued in the Golden Valley JV but as a diluting partner and management was passed to Anaconda.

In 1985 Anaconda's owner ARCO announced the termination of its mineral involvement world-wide. It offered its Woolwonga interests to its employees, who took up the offer, and backed them and other projects into an existing listed company, Dominion Mining Ltd which had, the previous year, acquired the Cosmo Howley prospect.

A range of small gold deposits were located and drilled by Geopeko but were deemed to be too small for PWL and all were sold.

In 1984 Darwin base was closed. The few Quest prospects outside the JV were farmed out. Stalled tenement applications were retained but Munmarlary records were shipped to Perth for safe-keeping and the building rented out. Bob Adams, supervising geologist retrenched at the time of closure, held a retainer from

Geopeko and occupied some of the office space for several years. Following the decision in 1992 to commence base metal search in the Carpentaria Basin, Darwin base was reopened with Terry Hoschke in charge.

Diamonds

The discovery of diamonds at Argyle in 1979, and early reports of its exceptional grade, were the catalyst for Geopeko to join the ranks of active diamond exploration. Exploration licences were applied for within the Auvergne 1:250,000 Sheet in the Victoria River region to generate primary anomalies for field testing and indicator mineral sampling. A magnetic and radiometric airborne survey was commissioned and a sampling campaign completed, with no encouraging results. Field operations and logistics were managed from the Darwin office, and a small-scale diamond laboratory was established in Perth to treat drainage samples and to ensure strict confidentiality of results. Magnetic and heavy media separation as well as mineral logging was undertaken at the laboratory, resulting in rapid and cost-effective processing. The commitment to diamonds was short lived, however, lasting from 1981 to early 1982, with exploration efforts confined to the Northern Territory, and allowed investigation of only one project.

QUEENSLAND

North-west Queensland

Reconnaissance in Gunpowder Creek area of North Western Queensland in the early sixties, which included a visit to the Mammoth (later called Gunpowder) mine, indicated a geological setting similar to that of Mt Isa. Two Authorities to Prospect covering an area of 30 by 10 km north, west, and south-west of the

Mammoth were applied for and granted in 1965. Air photos at a scale of 1:48,000 were obtained and a photo interpretation study of the geology undertaken before field work commenced. The study was a very rewarding one. The rocks were very well exposed and the rock units and structure easily identified. It confirmed a similar rock sequence to that at Mt Isa and a similar geological history with movement on fault blocks during deposition producing thicker units in the grabens and units thin or absent on the horsts. It reinforced the view that the area had copper potential but it could offer little towards indicating where to look. Field work checked out the few prominent "lode-like" outcrops (mainly iron-rich silicified stromatolite reefs) but most field work was concentrated on investigation of known copper occurrences.

One prospect, the Big Bend, ranked high enough for a drill hole. It recorded little in the target area but intersected a couple of metres of rich secondary copper in a fault zone deeper in the hole. The intersection was directly below the bed of a major creek and, wisely, the decision was made not to chase such a small prize in such a difficult place. After some months the conclusion was reached that to find a major copper deposit in the area Geopeko would need to apply sophisticated geochemical techniques, deeper penetrating geophysics, be prepared to drill deep holes for geological and geochemical information and, above all, be prepared to persist for some years. Without a nearby mine needing ore to prolong its life or nearby infrastructure to support the effort, such an exploration programme could not be substantiated and Geopeko withdrew in 1966.

Mt Morgan

Following the takeover of Mt Morgan Limited a Geopeko base was set up at Mt Morgan to provide mine services, the Mt Morgan geologists were taken into Geopeko, and an exploration campaign was developed under the stewardship of Ted Brennan, who had been running Mt Morgan exploration in the NT and had moved from Mt Bundey.

When it was acquired the Mt Morgan mine had only a few years of ore reserves remaining, with no obvious prospect of adding to them, so the principal effort was directed at exploration. An application for a very large authority to prospect, 508 M, was lodged over an area of 500 km² to the south-east and north-west of the mine in May 1968[21] and a major regional mapping and stream sampling program was undertaken, building on some very good work done during an earlier program by Consolidated Zinc Pty Ltd in joint venture with Mt Morgan. In 1967 Mount Morgan Limited geologists, led by Tony Hope, had recognised porphyry style copper and molybdenum mineralisation at Struck Oil, some 5km east of the mine, where mapping and soil sampling was subsequently carried out. Following the Peko takeover Geopeko ran a drilling program there.

An INPUT-EM survey was carried out over the license area but failed to find any conductive mineralization. Aeromagnetic surveys also failed to yield targets of significance. A regional mapping program by Alex Taube defined acid volcanic rocks which could host another Mt Morgan-type deposit. Ross Large, having completed a PhD on the Juno mine in Tennant Creek, was finalising

---

[21] A to P 508 remained in force for many years, establishing a record for the longest title held in Queensland at the time, and the reason was the quality of the continuing exploration recognised by the Queensland Mines Department including continuing mineral discoveries by Geopeko geologists and geophysicists led by Alex Taube and later in joint venture with Gold Fields Exploration Pty Limited.

a post-doctoral fellowship on volcanic massive sulphide deposits in Canada and was posted to Mt Morgan. He began mapping at the Ajax mine 25 km SE of Mt Morgan where he recognized the mine as a small volcanic massive sulphide deposit, and in short order discovered the Omo, Fab, Drive, and Grillo alteration zones with similar affinities.

Work continued over the Dee Range belt of rocks where major projects were found, including UNMC, which had some 2.5 km of stratabound volcanic massive sulphide mineralization. Most were of too low a grade at the time to support production. Other prospects identified including Mt Dick, Mt Alexander, and Raspberry Creek, and major regional geochemical anomalies were investigated at Fern Hills and Upper Don.

At one stage a rise in the copper price resulted in the re-examination of the Moonmera prospect. The deposit, a classical porphyry copper, had been discovered some years earlier by the previous joint venture, but proved to be uneconomic at the time.

In the late 1970s RGC Exploration elected to farm in to the Mt Morgan exploration areas and this provided a significant boost to funding for several years. Despite some rigorous exploration and encouraging results, none of these prospects yielded deposits of the size and grade required for the long-term continuation of the Mt Morgan mine, with only a small tonnage of near surface ore mined at Moonmera.

Mt Chalmers

In 1975 the Mt Chalmers mine leases, approximately 90 km north-east of Mt Morgan, were offered to Peko by Mt Isa Mines, initially as a JV but later as an outright purchase. The mine had been worked underground from the turn of the century to 1914, producing some 434,900 tonnes of copper-gold ore, so the eyes ostensibly had been picked out of it. But Geopeko and the Mt Morgan mining engineers found that there was a substantial tonnage of ore remaining that could be mined by open cut. An initial drilling program discovered the West Lode and added a modest tonnage of high grade ore which had not been worked by the earlier operators. The ore was mined by open cut and shipped by train to Mt Morgan, then blended with ore from Mt Morgan, which optimized recoveries for both.

The mine came on stream in June 1979 at the time of high commodity prices, which did not last the full three years of operation. It closed in May 1982. The Mt Morgan open cut had been closed in July 1981.

Meanwhile a major regional mapping and geophysical surveying program had been carried out over the whole of the Berserker Graben, host to the mineralisation in the region. When both of the open cuts at Mt Morgan and Mt Chalmers finished, the new Mt Morgan flash smelter continued to operate using concentrates brought by truck from Tennant Creek. Mt Morgan operations finally closed in January 1990. The discovery of the West Lode at Mt Chalmers had allowed mining, milling, and smelting operations to continue at Mt Morgan while the construction of a tailings retreatment plant was completed and commissioned.

In 1983 Geopeko decided to dilute its interest in the Mt Morgan joint venture. RGC became managers and took over the office. Exploration continued but with Geopeko diluting until 1990, when its remaining interest was sold to Elders Resources.

Townsville

A base was established in 1974 in Townsville with Warwick Maehl in charge. Exploration began over a range of provinces from Ravenswood and Charters Towers to Cape York. Targets initially were 'porphyry' copper-gold and volcanic massive sediment-associated base metal deposits. In due course a project west of Greenvale was selected, leading to the discovery of Surveyor One, a complex series of high-grade lead, zinc, and copper deposits. A failure to secure the entire belt left part of the field open to Mt Isa Mines who found an adjoining deposit. The two were later brought into production as the Surveyor/Balcooma Mine by Kagara Zinc Limited and were still operating in 2016.

A long campaign in the Irvinebank tin fields failed to bring success. It had been noted that magnetite was associated with the tin lodes there, including in mines that had been worked, such as the Vulcan. In September 1976 an area of 376 km^2 was flown by Geometrics with magnetometry and radiometry. From then to February 1979 there was follow-up exploration. Several mining leases were pegged and drilled. Success eluded the program and the project was shut down in 1979.

One of the more innovative projects was a drilling program in the Bloomfield River alluvial deposits from a barge brought in from Cairns. The target was alluvial tin below the Bloomfield Falls, thought to have been derived from the upstream China Camp field. However he grades obtained were insufficient to proceed.

On the departure of Maehl for New Zealand and with budget cuts looming the base was closed in 1982.

Brisbane

Andy Browne had set up a small office in Brisbane in mid-1982 to seek opportunities remote from Mt Morgan that had been neglected following the closure of the Townsville office.

The main focus for exploration initially was on epithermal gold in Carboniferous volcanics of the northern Drummond Basin of Queensland. The discoveries of Pajingo and then Wirralie and Yandan indicated the prospectivity of this region. Geopeko had several JV prospects with Menzies Gold namely Bimurra (near Mt Coolon) and Mt Violet (near Clermont). The Bimurra mineralisation was associated with Carboniferous rhyolitic volcaniclastic rocks and flows as well as several large rhyolitic masses. Bimurra mineralisation was particularly interesting with classical low sulphidation chalcedonic veining and platy or lattice quartz (silica replacement of bladed calcite). Siliceous sinters were also common at many of the prospects around Mt Coolon including Bimurra. The importance of sinters to exploration for epithermal ore deposits is that they provide tangible and unambiguous evidence of a palaeosurface. These sinters, although Carboniferous in age, looked very similar to sinters forming today (eg Champagne Pool in New Zealand). Unfortunately, no economic deposits were found during this exploration phase.

The discovery of Mt Rawdon (total endowment 2.51M ozs) hosted within the Triassic Aranbanga Volcanics (felsic volcaniclastic breccias) started a Geopeko focus on other similar aged felsic volcanic centres in south-east Queensland. The Rosehall vein system was discovered by Geopeko during several geological traverses across a volcanic complex near Mundubbera. Several pits were discovered on an approximately 2km long polymetallic single vein system (~1m wide). The prospect was systematically drilled but gold grades were too low to warrant further exploration. Kilkivan was a similar Triassic age epithermal mercury-gold historical mining area that was also explored for several seasons without discovering any economic mineralisation. The Barambah prospect with both epithermal Au and Ag as well as porphyry copper-gold was explored for several years. It was particularly interesting because its Cu:Au ratio was 1:1, that is, similar to the Goonumbla copper-gold ratio. However, drilling indicated that there was insufficient mineralisation to warrant further work.

With the sale of Rutile and Zircon Mines Geopeko had lost its association with heavy mineral ('HM') sand exploration and mining so it was decided to recommence HM exploration from the Brisbane office. As the exploration team had no experience in HM exploration, initially it was a giant learning process to understand the fundamental characteristics such as the importance of host rock provenance and the location of trap sites along the coastal plains. Ideally the target deposit should contain rutile, zircon, ilmenite, and leucoxene. A literature search highlighted that to generate rutile and zircon through weathering a certain metamorphic provenance was required. Therefore, geological maps showing metamorphic and stratigraphic and compositional information should be useful as prospecting tools for placer deposits. Consequently an exploration strategy emerged for use in Australia or overseas where one could choose favourable HM generating host rocks (eg granulitic pelitic sequences) then study the adjacent (if any) coastal plains for favourable trap sites using aerial photointerpretation.

A literature search in 1989 of Qld and northern NSW coastlines indicated that few areas remained available for exploration due to development of towns and fishing communities and National Parks and Recreation Reserves. Consequently the study focussed on more remote areas and one area of interest was Cape York. There the

Dargalong Metamorphics were of almandine amphibolite grade and prospectors had identified kyanite, monazite, and sillimanite from stream sediment sampling. A small HM deposit (Urquhart Point) was known near Weipa on the west side of the Cape. A literature study of previous exploration had shown that numerous explorers had looked for HM deposits along the eastern shoreline of Cape York and had identified rutile, zircon, ilmenite, and leucoxene, but in uneconomic quantities.

The team then pondered that perhaps the current Great Barrier Reef was high enough to restrict the wave and current action needed to form an economic HM deposit. But what if, during the Pleistocene, the reef level was lower and/or the sea level was higher allowing more wave and current action to concentrate heavy minerals. To preserve such deposits from subsequent erosion they would have to be situated inland. Aerial photos were obtained and in the first batch, covering an area along Princess Charlotte Bay 36 kms east-north-east of Coen, a large inland sand body was recognised immediately and became known as the Colmer Point HM discovery. The sand body (20km long and up to 3km wide) was mapped by the BMR as a residual sand mass derived from weathering of Mesozoic sandstones. The linear shoreline on the eastern side of the dune was mapped by the BMR as a fault. The dune was up to 80m thick and up to 100m above present sea level.

Geologist Graham Lee (Peter Stitt and Associates) provided guidance on the drilling, sampling, and analytical programs necessary to produce a resource estimate. The drilling and sampling was a logistical challenge, the prospect being approximately 600 km north of Cairns. A gas-fired drying oven was purchased so that the drilling sand samples could be dried and the size reduced by riffle splitting to minimise transport costs. Drilling indicated that the sand system was a low grade palaeodune and beach ridge deposit. Unfortunately no higher grade strand mineralisation was discovered below the dune and the deposit was identified as a large low grade system. An Inferred resource of 600,000,000 tonnes @ 0.68% ilmenite, 0.17% zircon and 0.06% rutile was determined from the drilling.

Other prospective areas to the north near the Lockhart River Aboriginal settlement were applied for but access was denied on archaeological and anthropological grounds. These areas were also partially covered in rainforest, creating environmental concerns. For those reasons, as well as the low grade nature of the deposit, the tenure lapsed.

Another HM target was Clinton Lowland within the Shoalwater Bay Military Training Area on the central coast of Queensland. The area was chosen from Landsat imagery by consulting photogeologist Tim Wilson because of prominent banded (strand-like) features in the area, and an ancient shoreline shape analogous to that at the Eneabba HM deposit in WA. Conservationists and the Military opposed the granting of tenure. A brief hand-auguring program was finally allowed but results were not encouraging enough to fight for tenure.

Other heavy mineral sand opportunities were identified on King Island and in the Murray Basin but were terminated when Aubrey Paverd took over as the North Limited exploration manager.

Mt Isa

Early in 1992 Geopeko decided that a move into base metal exploration in north-west Queensland was warranted. A projected move in the mid-eighties had been thwarted by budget cuts, continuing low base metal prices, and PWL's reluctance to explore for anything other than gold. Andy Browne had been geared up for another attempt in the late eighties, from Brisbane, but when CRA discovered Century available exploration tenure disappeared overnight. Colin Sinclair recommended a separate base in Mt Isa and Ian Mathison was given the job of setting it up and exploring the region.

Targets included zinc projects in the western succession, including Pegmont-type metasomatised deposits, copper-gold-oxide deposits such as Ernest Henry, and stratabound base metals in the Borroloola area. In 1995 the base was closed and the staff were transferred to Brisbane.

TASMANIA

King Island

As described in a previous chapter Geopeko's main focus following the acquisition of King Island Scheelite was to upgrade the mining operation, but at the time that company also held exploration tenements that covered the greater part of the island. Virtually no exploration had been done outside the area around the Grassy Granite and there was no comprehensive geological map. A team was posted from NSW, where they had been working on an offer for the Attunga tungsten deposits which PWL had declined to take up, and began the geological mapping, work that was essentially completed by the end of 1972.

The geophysicists applied gravimetric and magnetometric techniques to assist in the identification of structures and settings favourable for scheelite mineralisation. That assisted in the definition of the Grassy River Fault, which was thought to be one of the controlling structures for the scheelite mineralisation. Gravity surveys at Bold Head, across the contact between the Bold Head

Granite and the mine series rocks, provided useful information about the slope of that contact and consequently the thickness of the mine series rocks that host the mineralisation. As mining of the Dolphin deposits extended below sea level there was concern that workings might intersect the Grassy River Fault with the possible ingress of seawater. The geophysicists organised a crude marine magnetometer survey, consisting of a proton precession sensor in a (wooden) rowing boat towed 50 metres behind one of the many King Island fishing boats. Bob Richardson sat in the rear of the boat taking readings, recording into a tape recorder, and getting very seasick. Position control was achieved from two survey stations on the shore making timed measurements of bearings to the vessel. One complete total force traverse across the fault was measured and it was possible to model this and calculate with reasonable accuracy the position of the fault and its distance from the workings.

By 1981 tungsten production was going well but the uncontrolled export of wolfram from China into the Western market was starting to put pressure on world supply. In November 1982 the operation needed to be slowed down and one shift was dropped from production. In 1990 production ceased.

In addition to scheelite, King Island had also produced rutile from a small beach sand mining operation near Naracoopa on the east coast of the island, so rigorous exploration was begun of the multiple strand lines around the island. Although no new orebodies were discovered an excellent understanding of the regional geology of the island was achieved, some new scheelite deposits were located, and unusual tungsten, tin, and heavy mineral sand occurrences were investigated for their economic prospectivity, although none proved to be worthy of production at the time. A potentially economic resource of mineral sands was delineated at Naracoopa, but the Tasmanian Parks and Wildlife extended a nature reserve to cut off access to the second half of the resource and it did not proceed. It has since been mined.

Devonport

In 1975 exploration began on the Tasmanian mainland, run at first from King Island but from 1977 from a base set up in Devonport with Murray Rogers in charge. Following the practice from other bases "Voyager" was selected as the series name.

The initial target was the Mt Read Group, said to be one of the most prospective geological provinces for massive sulphide-hosted deposits in the world. Exploration tenements were obtained and were explored at Loongana and Black Bluff in the central north of the State and at Birch Inlet and Elliott Bay, the tenements covering virtually all of the southern extension of the Mt Read Volcanic Belt south of Macquarie Harbour. Targets in both regions were base metals, although geochemical results in the southern part of the Elliott Bay tenement suggested at one stage that discrete gold mineralization might be present. The gold anomalies proved spurious, however, and were attributed to the surface concentration of gold in the unusual weathering conditions of Tasmania's south-west.

Field seasons were short (Christmas to Easter) with access for equipment and supplies by barge from Strahan to the south-eastern end of Macquarie Harbour and then by Bombardier-tracked vehicles along an existing track. To avoid undue ground disturbance in the environmentally sensitive "button grass" country, in later years helicopters were employed extensively.

Over the next few summer seasons basic ground exploration was carried out, including geochemical and geophysical programs.

A Jacro diamond drill was mounted on one of the tracked vehicles and was used to obtain C-horizon soil samples for geochemical analysis and geological mapping. Short core holes were also drilled. Several prospects were located that were anomalous in both gold and lead-zinc.

Geopeko exploration continued in this area until, in 1982, the first of the big budget cutbacks of the eighties, at which time the project was farmed out to Aquitaine. Aquitaine withdrew in 1984 and the tenements were allowed to lapse. The exploration over the decade had been extensive and thorough. The geology had been shown to be very like that of the very productive Mt Read belt, and massive sulphides had been found at several prospects, but further exploration found no extensive mineralization.

Base metal exploration also continued in the north-eastern segment of the Mt Read belt, the most interesting and frustrating work being at Macintosh East. In this joint venture with CRA an airborne EM survey revealed one very strong anomaly among weaker ones. The strong anomaly, Anomaly 13, turned out to be about 200m inside the Cradle Mountain National Park, which in this area embraced a small corner of Mt Read Volcanics. The government was approached with the offer to purchase and donate some appropriate high country to the park in exchange for the excision of the small Anomaly 13 block. The approach was received with some warmth but political sensitivity meant that negotiations dragged on into the eighties, well beyond Geopeko's withdrawal from Tasmania. A change of government then swiftly killed the proposal.

Loss of the Elliot Bay JV partner in 1983 and failure to find a replacement had meant that there had been no field work during the December '83 to April '84 field season. Consequently fuel, buildings, and supplies from early 1983 were still at the Wart Hill campsite. With the decision to surrender the tenements there was a need to recover these to avoid damage to the environment. Following discussions with the Tasmanian Mines Department it was agreed that Geopeko would be allowed access during the late '84 early '85 season. Everything had to be transported overland via the Moores Valley route to Macquarie Harbour and then by ferry to Strahan.

Interest in tin and tungsten in the late seventies led to the establishment of a second exploration campaign, in north-western Tasmania. The targets were tungsten-bearing analogues of King Island and both stratiform and stockwork tin deposits, already represented in the region at Renison Bell and Aberfoyle respectively. Exploration was undertaken at Balfour, Pieman Heads, and Granville, among other areas, with drilling revealing mineralization but no ore grades. By 1982 this project consisted of the Rocky Cape EL, joint ventured with CRA, and the adjacent Montague EL. Project review associated with the 1984 budget cut ranked the project too low to be funded further and Geopeko closed the Devonport office and withdrew from Tasmania.

NEW SOUTH WALES

Coal Group

In 1971 a coal group had been set up in Newcastle with the aim of expanding PWL's coal assets. The Newcastle Wallsend Coal Company (NWCC) and its associates, J&A Brown and Abermain Seaham Collieries Limited, had been mining coal in the Hunter region since the mid 1800's. Diminishing reserves and a change in the coal market from local use to export meant that exploration had to be stepped up. Developments in mining techniques also meant the deeper mining was now possible.

Initially Paul le Messurier moved from King Island to Newcastle to carry out this work. Later Ron Lees transferred there to staff a Joint Venture between NWCC and Coal & Allied at Tahmoor south of Sydney, and at Jerrys Plains in the Hunter Valley. Significant resources of potentially underground mineable coal were defined in both areas, and the Geopeko team moved on to other bases. The Tahmoor resources were later mined by other companies.

Parkes

PWL had had mineral interests in New South Wales since the merger with Newcastle-Wallsend, including a holding in a mineral sands operation with Rutile and Zircon Mines, and had also completed an investigation of the tungsten deposits at Attunga which were on offer, but had declined to acquire them. With the establishment of a Sydney office the push to explore in NSW grew.

A number of reconnaissances by Elliston, Richardson, and others, including Kim Wright, concluded that the Lachlan Fold Belt had substantial potential for the use of geophysics and Geopeko's growing understanding of ore genesis to trigger the discovery of new deposits. In 1972 Parkes was selected as a suitable location for a base and Gary Jones was commissioned as geologist-in-charge. He chose the series name 'Endeavour' and it was not long before his team had several promising prospects identified. Early exploration success came with the discovery, in 1974, of a small lead-zinc skarn at Endeavour 7. It had a clear-cut magnetic, electrical, and geochemical expression. Magnetometric and self-potential surveys were completed, both of which showed well-formed concomitant anomalies. Drilling ultimately proved up approximately two million tonnes of ore grade lead and zinc mineralisation.

This result affirmed the validity of the decision to explore, and was followed up with an expanded program of regional exploration in the Goonumbla District north-west of Parkes that, in 1977, resulted in the discovery of the first of four major porphyry copper-gold deposits that comprise what is now the Northparkes Mine. Over the ensuing three years, to 1980, several more concealed porphyry deposits were discovered, including those at Endeavour 22, Endeavour 27, Endeavour 28, and Endeavour 26N. The first two, E22 and E27, were later mined by open pit, firstly for their supergene gold cap and then for supergene and primary copper. Drilling showed that Endeavour 28 was a large sub-economic system floored by a low angle thrust, but systematic RAB drilling to the south resulted in the discovery of Endeavour 26 North. This deposit contributed most of the ore in the first phase of mining. In 1992, over a decade later, Endeavour 48, a blind orebody, mid-way between E22-E27 and E26N, was found.

Late in 1981 the Board decided to farm out some equity in the Goonumbla project. Dr Ernest Miller was given the job of Development Manager on the project and Kim Bayliss the responsibility for preparing tender documentation and assessing the potential partners.

In 1982 Chevron joined Geopeko in the joint venture. It seconded several of its professionals to the team and provided a Technical Co-ordinator to oversee the program. Jones in particular, but Geopeko as a whole, found themselves struggling to cope with an exploration culture totally alien to its operating philosophy. Fundamental to Geopeko's way of operation was the belief that the project geoscientists were the best equipped to plan the programs, to interpret the results of their fieldwork, and to find the orebodies. The role of Head Office was to set the strategy, provide the funds and encouragement, and to critically monitor progress. Chevron regarded the field people as data collectors for experienced geoscientists (mostly in the US) to interpret results and plan each step of the program.

Jones resisted strongly to the added workload of detailed reporting and preparation of detailed proposals, objected to the delays in approvals and resulting inefficiency, and he smarted at the removal of flexibility that had allowed him previously to get on with the job that he had done so well. There were some fiery sessions between him and the Chevron representatives at the Operating Committee table. He had many wins, not the least being able to convince Chevron that drilling of a vertical pipe-like deposit is best done with angle drill holes, not vertical ones as was the practise for porphyry deposits in the south-west USA.

The Goonumbla discoveries provided Geopeko with an exploration advantage over its competitors and it moved quickly to capitalise on that opportunity. New areas with similar geological and geophysical characteristics were sought elsewhere in the Lachlan Fold Belt. A kick-off meeting of what was to become the Lachlan Project involving the Goonumbla team, Brian Williams and Andy Browne of the Project Development team, and Bob Richardson and Geoff Sherrington representing geophysics and geochemistry respectively, was held in Parkes. Jones described the characteristics of the deposits and their geological setting. Similarly, the geophysical characteristics of Goonumbla district and the geochemistry of the host rocks were discussed. A short memo summarising the first few months progress was prepared by Brian Williams in May 1980. This memo and its ten area selection parameters formed the basis of an extensive search, commencing in 1980, that eventually covered most of south-eastern N.S.W and parts of Victoria.

The initial study identified three areas to the north of Parkes, the Lake Cowal area in the West Wyalong district, and a number to the east in the well-prospected part of the Lachlan Geosyncline. Lake Cowal ranked highly and was secured by a large block of 6 EL's. Smaller blocks were taken over several of the other areas. Exploration at Lake Cowal was hampered by soil cover up to 100 metres deep and an almost total lack of outcrop. But despite the difficulties in drilling deep scout holes, early exploration results including the discovery of Endeavour 39, a low grade porphyry deposit containing several billion tonnes at 0.2% copper, offered strong encouragement for the discovery of a repetition of the Goonumbla style mineralisation. Geopeko persisted with this program throughout the early 1980s, under the guidance of Mick Love, and achieved success with the discovery of the major Endeavour 42 (Lake Cowal) porphyry-style gold deposit in 1988.

In 1979 Geoff Sherrington relocated an XRF instrument, no longer needed in Darwin, to the Gordon office. The Geochemical team at Gordon offered a free scanning service to all the exploration bases and the Lachlan Project took advantage of this service routinely sending suites of samples to be scanned. In 1980 Mesozoic trachyte from the Toongai area south of Dubbo was found to be strongly anomalous in Niobium. In 1981, detailed mapping and sampling by Wayne Hoy defined two prospects, Plens (Endeavour38) and Hylands. Six short holes were drilled at Hylands and indicated a large, potentially economic Niobium resource in excess of 70Mt. Later, a scoping study had difficulty in identifying the Niobium bearing mineral but indicated very difficult metallurgy and a vulnerability to competition from the Araxa Niobium deposit in Brazil. It was concluded that the project could not be brought into production in the short term and all tenements were abandoned in early 1983.

Alkane Resources recognised the projects potential and

acquired the ground almost immediately. They noted that Hylands also contained Zircon and Rare Earth Elements and that they might be economically significant. It has taken almost 35 years to achieve acceptable recoveries, but Alkane is confident that they will commence production at the Dubbo Zirconia Project in 2018.

Chevron had started farming into the Lachlan project in 1984 but withdrew well before the Endeavour 42 discovery. It also withdrew from the Goonumbla project in early 1987 leaving PWL and then North, when it took over, to take the project, by then renamed Northparkes, through to feasibility in 1992. At about this time a 20 percent share in the project was sold to Sumitomo, linked to a copper off-take agreement.

Following the withdrawal of Chevron, Geopeko resumed Goonumbla project exploration, aimed at first at finding epithermal gold resources to complement the supergene gold which would be the first ore to be produced from the porphyry deposits. The Endeavour 44 deposit was found adjacent to the Endeavour 7 skarn, and a farm-in was negotiated with Alkane over the remaining deposit beneath the old Peak Hill mine north of Parkes. Both were extensively drilled but the Endeavour 44 resource proved too small and the Peak Hill ore too metallurgically challenging, and they were dropped.

Late in the eighties when the Northparkes project was well advanced towards feasibility, an MMD-imposed embargo on copper exploration was lifted and Geopeko was encouraged to resume porphyry search. Reworking of the extensive database accumulated over the previous two decades provided further targets and led to the discovery of the Endeavour 48 deposit in 1992.

The Northparkes mine was commissioned in 1994, based on a resource of 86 million tonnes of ore averaging 1.27% copper and

0.66g/t gold. That included only the upper part of the Endeavour 48 deposit, which was still being drilled. Drilling of the Lake Cowal resource delineated 50 million tonnes of ore averaging 1.5g/t gold (2.4 million oz) by 1996, when the project reached the feasibility stage. The NSW Government refused development approval in 1996 but the decision was reversed three years later. Development stalled, however, and the deposit was sold to Homestake in 2001. Barrick, which had taken over Homestake, began mining in 2005.

Peak Hill

In the mid-eighties Geopeko, from its Parkes base, was seeking shallow gold resources which could add feed to a start-up gold treatment plant and improve the overall economics of the Goonumbla copper-gold project.

In August 1986, the Alkane-Paringa-Molopo-MMS joint venture advertised its Peak Hill, NSW, gold prospect and the London Victoria prospect on the outskirts of Parkes township for farm-in. Both had been mined to relatively shallow depths late in the nineteenth century. The Alkane joint venturers had undertaken exploration to prove up extensions to the original resource which might be extracted profitably using modern technology. Geopeko expressed an interest as did a number of other companies including BHP Gold and Shell. Although both deposits were of interest to Peko, being gold resources which might supply early ore feed to a Goonumbla operation, Peak Hill, the larger at 2 million tonnes averaging 2.2g/t gold, had the greater appeal. BHP Gold moved fast and secured the London Victoria, a quartz vein hosted deposit which presented little difficulty and relatively little cost to bring to production. Geopeko secured Peak Hill, an epithermal deposit with most of the gold associated with disseminated pyrite within a broad zone of clay-altered rock, in mid-December of that year. Assessment of existing exploration data followed and drilling to confirm the resource and to obtain samples for metallurgical test work began in mid-January.

Exploration by O'Neill and his Goonumbla team was essentially finished by mid-year with the resource confirmed and slightly increased. Results of metallurgical test work had not been so favourable. Clay content in the ore was causing problems and gold recovery was very low. Further test work continued through 1987 and into 1988. In February, Geopeko sampled the old tailings in the hope that their addition to the resource might supply some easily extractable gold and improve the financial model. A month later when Geopeko's projects Australia-wide were presented to the North Board following the company take-over, it was stated that gold recovery from Peak Hill ore could not be improved to a satisfactory level and viability of mining looked remote. Nevertheless, the Alkane JV was approached with a request for renegotiation of more favourable terms to assist Peko in trying to make the project financially viable. The request was declined and Geopeko withdrew from the JV.

Ashton Mining followed Geopeko, reaching agreement with the Alkane partners to continue assessment. By early 1990 it was reported to have obtained reasonable gold recoveries and was doing mine design for a heap leach operation. But by February 1991 it was pulling out because the proposed open pit extended too close to houses in the town. Alkane then decided to go it alone. Within a year it had completed a feasibility study, was reporting gold recovery of 77%, and had most of the town on side for mining. Getting final approvals took some time but mining commenced and the resource was successfully extracted between then and 2001.

Cobar

The Cobar exploration base was opened by Leigh Schmidt for Norgold Limited in 1988 to apply some newer thinking to a field which had had a long exploration history and was very familiar to him. He had been a key member of the EZ team which had found the Elura copper-lead-zinc-silver deposit there 15 years earlier.

The target was traditional Cobar-style gold and base metal deposits, on faults or in anticlines above blind faults. There was to be some emphasis on gold, reflecting Norgold's charter, and perhaps also the recent success of The Peak mine. Most of the Cobar deposits contain magnetite and/or pyrrhotite. Consequently aeromagnetometry had been the main regional exploration tool.

Schmidt proposed a new geochemical approach that would provide a method of focusing on non-magnetic deposits and also on those weakly magnetic ones "hidden" within the noisy magnetometric pattern for which the Cobar district is renowned. The exploration, comprising detailed sampling along drainage lines located by shallow drilling, was akin to the prospector's loaming up a stream, but more sophisticated and much more expensive. The program which, at its peak, encompassed 34 Exploration Licences, was known as the Cobar Supergroup Project.

Two other projects were run or proposed from the Cobar Base. One was a limited search for base metal deposits within the basement rocks to the east of Cobar, using magnetometry mainly. The other, named the Trans Darling Project, was aimed at locating porphyry or Cobar style mineralization beneath the Great Artesian Basin on the north side of the Darling Lineament north-west of Cobar. The Cobar Supergroup Project was well underway by the time Norgold was absorbed back into North and Geopeko became the explorer for the North group in February 1990. Results were not encouraging. Anomalies were very subdued and attempts to locate their source had either failed or identified only weak mineralization.

Things changed only a month or so later, however, when a siliceous floater picked up near McKinnons Tank during the stream sampling in the area recorded significant gold values. A prospecting exercise that followed found outcropping mineralization a few hundred metres to the south. Drilling, however, dampened enthusiasm and by July the mineralization was being considered to be limited in depth and probably only the result of supergene enrichment. A resource estimate in November identified a high grade body of 90,000 tonnes averaging 6.6g/t Au, surrounded by a low grade envelope of 1.5 million tonnes at 1.6g/t Au. A couple of months later there was a comment that the high grade section would not be big enough for a CIP treatment plant and probably should just be included within a heap leach operation. Drilling did continue through 1991 and into 1992. The resource reached 1.7 million tonnes averaging 1.6g/t Au. A Preliminary Feasibility Study indicated that mining would be viable, but North found the project too small to meet its development parameters. In May 1992 it was decided to seek a farm-in partner or buyer while drilling was completed and a mining lease applied for. It was sold to Burdekin Resources in 1993.

Early in 1993, the decision was taken to close the Cobar base. McKinnons had proved too small to be of interest to North and would be sold. The palaeo-drainage sampling targeting Cobar-type mineralization had failed to produce any significant anomalies, and grass-roots exploration for deeply buried mineralization along the Darling Lineament had proved too daunting.

WESTERN AUSTRALIA

Perth

In 1974 Brian Williams moved to Perth to open a base there, staffed to a great extent to begin with by those who had had to leave the Darwin and Jabiru operations because of the shut-down of uranium exploration.

Paterson Range

In 1975 Carr Boyd Minerals approached Geopeko for a Joint Venture in the Paterson Range Province and a base was established there to support a major exploration campaign. Over the next three years a substantial exploration program, comprising detailed mapping, geophysical surveys, and extensive geochemical programs, backed by scout drilling, was completed. It was shut down in 1978 for financial reasons, before any discoveries had been made.

Peak Hill

A year earlier Ray Twist, who had come from the Darwin base, suggested that some of the bif-associated gold deposits in the Murchison geological province were similar to those at Tennant Creek. That triggered an investigation that led to the development of the Peak Hill mine, a previously worked gold deposit selected for its potential as a shallow oxide resource which could be rapidly brought into production. There were a number of similar open pit mines being developed in Western Australia at the time, with the ore treated by Carbon-in-pulp ('CIP') extraction. MMD, however, considered it to be too low grade and didn't immediately see the potential for low cost treatment that CIP offered. There were efforts to sell it during 1982 but they failed. In mid 1983 a small group, Grants Patch Partners, later to become Grants Patch Mining, began treatment of the tailings under tribute, which led to a joint venture between Geopeko and Grants Patch with Geopeko diluting.

Funding continued to be withheld until 1985 when finally there was the realisation that Peak Hill might make a small mine. By this time Geopeko's equity had fallen to 50%. Resource drilling recommenced and the project was taken to feasibility in 1986 with an open pit reserve of 109,000 oz gold. Mining began in 1997. Grade proved better than had been predicted, as did gold recovery. Further ore was found nearby and ore from the Baxter's deposit a few kilometres away was brought to the plant before the mine closed in 2000, having produced over a half million ounces of gold.

Bangemall Basin

An offer by Amoco Minerals to look at the Bangemall Basin led to the formation of the Jillawarra JV and the discovery of the Abra base metal deposit. Amoco had for several years been exploring in the central part of the basin where there were some gossans anomalous in base metals. Geopeko's involvement was based on Lew Richardson's work on magnetic modelling. His expertise in this area was the reason for the invitation from Amoco to join the search. An aeromagnetic survey had produced unexpected results – several very large, intense, bulls-eye anomalies – in what would normally have been considered a magnetically quiescent set of sedimentary rocks. The magnetic source of the anomalies had to be at several hundred metre's depth but all Amoco's attempts to get an answer to what was the source through drilling had failed. This time it was Geopeko's exploration record at Tennant Creek which brought the American oil company seeking help. David Timms requested a meeting and showed the results from their Jillawarra project. The magnetic anomalies were just like those hosting mineralised ironstones at Tennant Creek only much larger, particularly one named Abra.

Abra was the first prospect drilled by Geopeko and the hole recorded hundreds of metres of lode material consisting of magnetite-hematite-carbonate-barytes with variable but mostly low grade lead, a little copper, low gold values in places, but virtually no zinc. There followed extensive drilling of Abra seeking higher grades, a gold orebody within the large mineral body, or the elusive zinc. Other prospects investigated within the joint venture area included Cadabra, TC, and TP. Geopeko flew aeromagnetic surveys in other parts of the Bangemall Basin outside the JV area seeking repetitions of the Jillawarra mineralization. Terry Ballinger and Julian Hanna were heavily involved. Geophysicist Andy Mutton, who had joined the team in 1982 when geophysical support had been decentralised, found himself occupied almost full time on this base metal search. These two projects survived the cuts of 1982, the first of Geopeko's Australia-wide budget cuts.

In 1984 Geopeko was told to dilute its interest. In September 1985 a farm-in partner for Jillawarra was found, SAMIM electing to earn a third through exploration spending. Within a year both SAMIM and Amoco had withdrawn. Geopeko continued search for the elusive zinc, drilling a number of deep holes over the next five or so years. An inferred resource of 56 million tonnes averaging 4.5% lead was estimated in 1991, too low a grade on lead alone to warrant firming up the resource, but Abra was held as an inactive prospect into the mid-nineties in the hope that there might be a technical breakthrough prompting further exploration.

Kalgoorlie

In 1982 it was decided to attempt to put together some large holdings on prospective ground in the Yilgarn Province, either by farm-in or by acquisition of the many smallholdings which typically formed a complex cover in the old mining centres. An office was set up in Kalgoorlie with Ray Twist and Terry Ballinger directing exploration. Larger blocks were needed in order to underwrite systematic grass-roots exploration and to have sufficient ground for mine development if a viable deposit was discovered. To Twist and Ballinger Kanowna was an obvious choice. The field, long since abandoned except for 'gouger' operations, had produced over a million ounces of gold, and from a range of different deposit types, with the inference that they had resulted from a substantial and long-lived flux of gold bearing fluids.

Twist set about finding tenement holders and negotiating deals with them. Unfortunately, David Gelattly of Delta Gold NL had the same idea at about the same time and Geopeko found itself competing with him. Gelattly got what was thought to be the prizes, the roots of the White Feather vein quartz deposit, a portion of the "deep lead" potential, and some unexplored ground to the southwest for Canyon Resources, a related company. Geopeko got the core of the 'deep leads', the Fitzroy, Cemetery, and QED leads, and a large portion of the Red Hill porphyry with its stockwork potential. Canyon was short of cash and, before it was absorbed by Delta, farmed out its south-west block to Geopeko with a fixed sum to be spent on exploration to earn 50% equity but a commitment to fund fully to feasibility if it proceeded further.

Both groups discovered shallow gold deposits, secondary deposits in the soil above small gold-bearing veins, while continuing to explore the extent of their tenement holdings. Delta proposed a jointly owned mill to treat these ores; but PWL was not enthusiastic. When the suggestion was changed to a 50:50 joint venture, however, there was agreement and terms were settled in late 1987. PWL insisted on being the miner, forcing Geopeko to cede exploration management to Delta. A week after the deal was concluded Ballinger received the results of Geopeko's most recent scout drilling phase. Two RAB holes recorded about 4g/t gold from primary rock in the base of the holes and were obviously the indicators of the presence of a deposit of substantial interest. Priority follow-up drilling should have been the normal response to such results but Delta's project geologist had other views.

Those two drill holes were the holes that led to the discovery of Kanowna Belle, but that was not confirmed for another two years when Delta returned to the discovery locality. Ballinger failed to get the credit for an excellent bit of exploration.

From the first test hole it was apparent that this was a significant discovery. Surface drilling continued to mid-1993. Mine development started well before this, based on open pit reserves. The resource down to 600m at the time of mine opening in November 1993 was estimated to be 22 million tonnes averaging 5.7g/t gold (a little over 4 million ounces).

In the early eighties more budget cuts had led to a review of operations in WA. Five projects were to be dropped, four retained. There would be dilution in the Jillawara JV if a partner could be found, and continued dilution at Peak Hill. On four more (Glengarry and three from Kanowna, UDAP, Delta, and Red Hill) it was undecided initially but the Kanowna projects were retained after further consideration. The Delta farm-in area contained Kanowna Belle, which would be developed some years later. The offices in Kalgoorlie remained open.

In the late eighties Geopeko concluded a farm-in with City Resources over a large block of ground in the Kalpini district about 80km north-east of Kalgoorlie. To Ballinger it offered promise because of the presence of old mines, carbonate alteration, conglomerates (indicating an extensional tectonic regime), and proximity to a major magnetic feature. City Resources had undertaken a regional soil geochemical sampling program and some detailed follow-up had produced anomalies that were ready for drill testing.

Geopeko tested each of the anomalies with RAB drilling, but in most of the cases did not locate a source. Drilling of one anomaly, however, intersected the silicified top of what proved to be a felsic andesite unit containing pyrite, arsenopyrite, and gold. In 1989 this became the Mayday North prospect. RAB drilling soon delineated gold mineralisation in what turned out to be a 5 to 10m thick supergene-enriched zone below 30m depth, extending 150m north-south and up to 70m east-west. Grades were of the order of 3g/t gold. Deeper drilling, however, showed that the primary mineralization was arsenopyrite and pyrite in volcanic rocks with gold grade about 1g/t, much too low to be of economic interest.

There was more deeper drilling, including two diamond core holes, but no high grade shoots were found, the best primary intersection being 25m averaging 2.2g/t gold. In June 1990 a resource of 127,000tonnes at 3.7g/t gold was estimated for the supergene material, extractable by open pit with a strip ratio of 16:1. This was much too small for a stand-alone CIP plant and the deposit was parked pending a decision either to mine and truck ore 30km to the Kanowna Belle plant, which by then looked a possibility, or to sell it for treatment through another local plant. Closer to Kanowna the Pigmy Pony prospect had geological features very similar to those at Kanowna Belle, anomalous gold, and geophysical anomalies. Extensive drilling located scattered ore grade intercepts but failed to delineate a commercially viable deposit.

In May 1988, in response to an offer from Delta, Geopeko undertook an assessment of the Granny Smith gold project. CSR had decided to withdraw from the project and Delta had pre-emptive rights to buy CSR's 50% interest in the joint venture. Delta was asking for financial support in exchange for equity in a new joint venture. The assessment revealed a well-defined and growing resource, potentially economic grades, and untested resource potential. It was strongly recommended that North negotiate with Delta but the Board declined. Delta moved on to do a deal with Placer and Granny Smith developed into a major gold operation.

The Platinum Saga

In 1979 Brian Williams and Geoff Sherrington had undertaken a brief review of platinum group metals ('PGM') potential in Australia. They concluded that Australia had potential for economic PGM deposits, recommended the east Kimberley and central Yilgarn layered mafic complexes as targets, and suggested the setting up of a task force to investigate the potential. Four years later Alex Taube became the 'one-man' taskforce. He produced a report which recommended the Panton layered ultramafic intrusive within the Lamboo Complex in the Kimberley as the most prospective. In April 1984 an office was opened in Kununurra, but the campaign was cut short by the budget cut which came into effect immediately afterwards. In 1986 work resumed and Geopeko acquired ground within the Lamboo Complex. Field work began in mid-1987 following the wet season but there followed a year of acrimonious negotiations with the State Government and the Kimberley Land Council triggered by an erroneous report that sacred sites had been disturbed. Geopeko finally regained access late in 1988 and exploration continued into 1989. Results were not encouraging and at year's end the company surrendered its licences.

Late in 1989 another layered mafic intrusion, the Windimurra Complex, 60km east of Mt Magnet in the state's Yilgarn Province, became available for farm-in from Pancontinental and Degussa. Windimurra had been on Taube's list and was a favourite of geologist John Bunting, who was running the program and considered it very similar to the Stillwater Complex in Oregon, which had a producing PGM mine. He believed that the Pancontinental-Degussa exploration had not tested low enough in the layered sequence. A farm-in was negotiated and exploration began in the 1990-91 budget year, targeting the largely non-outcropping lower part of the complex. There was little encouragement however and after a few months the project was brought to an end.

SOUTH AUSTRALIA

Adelaide

Following the closure of the Cobar Base in NSW there was a redistribution of junior staff. Leigh Schmidt was commissioned to set up an exploration base in Adelaide, focused mostly on base metal deposits in the Stuart Shelf and Curnamona Provinces. The office opened in March 1990.

Exploration on the Stuart Shelf had begun in April 1979, when it was a three-way JV between Seltrust, Mt Isa Mines (represented by Carpentaria Exploration Company – 'CEC'), and Sims Metal (a PWL subsidiary), with Seltrust managing. Its target was stratiform copper in the Stuart Shelf sediments. After the acquisition of Sims Geopeko was asked by PWL to represent the group in the Joint Venture. The stratiform copper search had located several thin relatively low grade deposits but the rising importance of Western Mining's Olympic Dam discovery soon deflected the search to the basement rocks. Seltrust withdrew within the year, leaving CEC as managers of a 50:50 JV with PWL. Prospects explored included Horse Well, Red Lake, Salt Creek, Portulaca Ridge, and Baker Dam. The best result from this early phase was a few metres assaying 2% Cu and 0.4g/t Au in hematite at a depth of over 750m at Emmie Bluff.

Another hole at Emmie Bluff in June 1984 gave some real encouragement. It intersected 80m of hematite averaging 0.7% copper, but gold levels were low at 0.2g/t. The plan was to pre-drill a follow-up hole and then seek a partner to ease the financial burden on CEC and Geopeko. A partner could not be found, however, and the follow-up hole in August 1985 recorded much the same result – 350m of haematitic rock below 750m depth but with the best copper assay being 2% and the best gold 0.4g/t. CEC expressed its wish to persist with the campaign and to drill holes on several more prospects.

In May 1986, the SA Mines Department advised that WMC was proposing to drop off a large amount of exploration ground which would be opened for tender by other explorers. CEC, on behalf of the JV, assessed the prospectivity of the areas and the JV approved the project in September. By September 1987 another hole was being drilled in the Emmie Bluff tenement. It intersected over 60m of hematite below 800m with the best intersection 18m averaging 0.9% Cu and 0.3g/t Au.

With copper price low and both explorers short of funds the mid-eighties were very quiet but the JV managed to hold on to its ground. Drilling of hole SAE4 late in 1987 recorded 18m at 1% Cu and 0.3g/t Au within a wider intersection of hematite mineralisation. Step-out holes over the next 2 years confirmed the presence of an Olympic Dam-like environment but the best intersection was 15m at 1.23%Cu.

Late in 1989, with copper price rising, MIM proposed a swing back to stratiform copper exploration. A number of holes were drilled in 1990 and 1991, targeting the Tapley Hill Formation, in the hope of defining a viable deposit in a small sub-basin. Intersections included 1.7m at 1.6% Cu, 6m at 1.6 % Cu, and 1.9m at 3.4% Cu but the resources were too small for mining to be considered.

In mid-1993 Schmidt suggested that Geopeko take over management and review the past work, but the project was dropped.

VICTORIA

Ballarat and Ararat

In July 1990 the company's Acquisitions Group had been in contact with Ballarat Goldfields Ltd ('BGF') regarding a possible farm-in to its East Ballarat project, and had contracted consultant Roy Cox to review the project before bringing it to the notice of Geopeko. Ballarat Goldfields had acquired tenure over the old East Ballarat field, which had produced over a million ounces of gold from fifteen mines before closure during World War I. The potential prize was another million ounces for every 300m of vertical depth, from ore that historically had had grades ranging from 7 g/t Au to 17g/t Au and averaging 10g/t Au.

After a visit Cox was questioned at length by Dr Miller on the nature of the ore and wall rocks. It was clearly established that the main ore source had been ladder veins, known locally as 'Leather Jackets', containing unevenly distributed nuggetty gold; and that all drilling could hope to do would be to find and establish the location and size of new veins. Resource definition would only come by establishing the gross dimensions of each lode and an estimation of the likely grade based on previous production. The group decided to make an offer to farm in to a 60% interest, and Geopeko was given the job of an on-site review.

The main responsibility fell to Wayne O'Neill who had recently accepted the position of Supervising Geologist – Victoria. He found from the drilling that the nuggetty nature of the quartz lodes was even more extreme than had been anticipated. Individual samples might assay from 60 g/t Au to 600 g/t Au, but all nearby samples might assay less than 1g/t. The leather jackets also seemed to be relatively small, up to a maximum of about 150,000t, making it necessary to have mine production from three or four of them at any one time to reach a desirable annual gold output. That in turn would put pressure on underground exploration and on mine development.

These problems were reported to Dr Miller but he indicated that they were no surprise to him (having previously worked in quartz vein gold mines in India). He felt that there was an adequate prize to be located over the length of the mineralization and that the risk was worth taking.

O'Neill moved his team into BGF's office in Ballarat, set up the voluminous historical data in a database, and used this to model the geology and limits of the mined orebodies with the aim of defining new targets. Drilling got under way in January 1991 on what was a two-year plan for Geopeko and MMD to bring the project to a decision point. By May there were four drills operating, all on a 5 day a week day shift only basis to keep noise within urban Ballarat at acceptable levels. Dewatering of the old mines was continuing, and re-collaring of one of the old shafts was being considered with the aim of getting underground to inspect the old workings when dewatering was completed. Bob Spark, BGF's Managing Director, had done a superb job on community relations and, through consultation, had set up efficient and low risk operations within the city. O'Neill was able to continue and expand on these with no community objection.

In September, MMD commissioned mining consultant Ted Davies to work with O'Neill to determine the range of capital and operation costs that would be needed to develop a target of a million ounces of gold in ore of grade greater than 6g/t. Ted was well known in PWL, having been Peko's Manager of Operations at Tennant Creek in the 70's and a senior manager in the Wallsend Coal operations.

His report in November was damning. He concluded that a grade of 15g/t was needed to make mining below 300m viable. History indicated that that was not achievable. BGF brought in other consultants and the economics looked more favourable the second time around, but North had been sufficiently spooked. By January 1992 the decision had been taken to withdraw from the joint venture. O'Neill and many in Geopeko were far from happy with the outcome. Geopeko had been led into making an intense and expensive effort on the project, somewhat against its gut feelings, only to learn that the effort was wasted and could have been avoided had the financial analysis been done at the outset. Following Geopeko's withdrawal from the joint venture Ballarat Goldfield Limited were able to secure other joint venture partners and to move the project to production. Underground mining was still going in 2015, including some development of the mineralisation identified during Geopeko's 1991 drilling.

O'Neill had started some other project work during the Ballarat campaign. The Stavely Volcanics were of interest because of the presence of andesitic volcanics and the fact that there was a significant regional magnetic and gravity ridge that suggested a possible southern extension of the Lachlan Geosyncline from NSW and therefore potential for the presence of porphyry copper mineralisation. On a reconnaissance trip O'Neill discovered a small sub-crop of altered quartz-veined intrusive, and immediately applied for a number of Exploration Licences.

With the closure of the Ballarat Project O'Neill moved to Ararat and redirected work to the Stavely Project. Literature search of previous exploration activity revealed information on the presence of copper mineralisation at a prospect known as Thursday's Gossan. Systematic RAB and RC drilling there by Geopeko led to the definition of a large (5kmx2km) low-grade porphyry copper system. Secondary chalcocite mineralisation was identified. The deposit proved to be too low grade to warrant further work however.[22]

Additional Exploration Licences were applied for over both the northern and southern extensions of the Stavely Belt. Two other mineralised intrusives were discovered to the south of Thursday's Gossan but in both cases copper grades were very low. To the north, in the Dimboola/Horsham area, the thickness of cover by Murray Basin sedimentary rocks was a concern.

Grassroots gold exploration was also carried out from the Ararat office. The target was Bendigo Style mineralisation under Murray Basin cover. It was recognised that whilst Ballarat East was not a suitable target, Bendigo with its historically higher grade and stronger continuity of mineralisation was a more attractive target. The presence of high levels of arsenic in ground waters surrounding both the Ballarat and Bendigo deposits led to a collaboration with CSIRO to test existing water bores to the north of Bendigo. Microgravity surveys were also used in an attempt to identify depth to basement and delineate paleo-drainage systems that might be amenable to geochemical sampling.

---

[22] Quarterly Report, Stavely Minerals, September 2015.

Changes in corporate strategy led to the cessation of work in Victoria and closure of the Ararat Office in 1995. Since then the Victorian Government has undertaken the major 'Gold Undercover' initiative to stimulate exploration for gold mineralisation concealed beneath Murray Basin cover to the north of Bendigo. Three separate gold occurrences have been located. Geoscience Australia and the Victorian Geological Survey are currently engaged in a major initiative to stimulate further exploration of the Stavely Volcanic Belt, particularly those areas under cover.

OVERSEAS EXPLORATION

In August 1990 the North Board requested that Geopeko give its view on overseas exploration. Geopeko had had its recommendations on exploration offshore refused by the PWL Board so many times that it had long since given up trying. This request therefore came as a real surprise. It looked firstly to regions where there was established potential for important discoveries, ones that Geopeko could hope to make through application of its experience and integrated approach to exploration, and secondly to countries where business, political, and social systems were not too far removed from those in Australia. The latter was to ease the pain of setting up but equally to reduce the perceived risk particularly for North's conservative Board.

The team settled on the United States as top priority, with Carlin-type gold deposits and south-western type copper porphyries the main targets. Warman International also had a significant presence in the US and was happy to offer assistance with getting established. Anxious to use the Parkes and Lake Cowal experience with alkaline intrusive-related porphyry copper-gold deposits to enter the search for "super porphyries" in SE Asia, Malaysia was nominated as second country. It was seen as having retained much of the British legal and commercial system, corruption was far from the problem reported in its neighbours, and sovereign risk appeared low.

At the end of 1991, Dr Miller retired. Malcolm Broomhead stepped into his position of Executive Director, Mining and Industrial, and almost immediately called for a review of exploration and exploration strategy as a familiarisation exercise. He was supportive of more offshore exploration and called for a broader review of potential.

Submissions were sought from staff, and a small list of consultants, on recommended regions or countries, and on the types of deposits to be targeted. Reviews or proposals were received from fifteen Geopeko staff and focused on South-east Asia, China, India, and Eastern Europe, but included contributions on Ireland, Spain, Zimbabwe, Canada, and Argentina. A summary of the proposals came out strongly for the Western Pacific and Eastern Europe.

Up to 1994 the parent board was risk averse and reluctant to commit in many cases, and serious overseas involvement was limited to New Caledonia, Chile, New Zealand, and the USA, although there were tentative forays into Malaysia and Indonesia. In the following six years leading up to Rio Tinto's take over of North, however, North had a major push to internationalise. This took North Exploration, the renamed Geopeko, into Indonesia, Europe and Africa.

USA

In June 1991, the Board gave its support to a move into the USA, stipulating that the Geopeko representative must be an experienced senior who was prepared to stay in the position for at least 3 years. It deferred a decision on Malaysia until more information had been obtained. Andy Browne was invited to lead the US team and accepted. The first moves to set up an operating company were made in October. North was not happy with the use of the name Geopeko and the company became North Mining Inc. It was a surprise to learn that the name Geopeko, and its record, were far from unknown and that North Mining meant nothing to our US counterparts. A fact-finding trip was made to Nevada and Denver in November and Denver was chosen as the best site for an office. In March 1992 Browne and Calder were in Denver finding office accommodation and housing. Mick Love joined as second-in-command a month or so later.

It took several months to settle the families, move into and equip the new office. Then followed an extensive series of field trips throughout the western U.S. Many of the major copper and gold mines were visited, contacts were established, and an understanding of the geology and mineralization of the western USA was developed. It was clear that if the company was to be successful in this mature well-explored terrain its exploration approach would have to be different. It would be a waste of time and money simply to do what everybody else was doing.

Most exploration at the time was for gold and tended to be confined to near-mine tenements in Nevada, with little or no real greenfields exploration. In addition, although the published geology seemed to be good, there was no governmental reporting system to preserve exploration results, and the publicly available geophysical data, mainly magnetics and gravity, were very old, broad-spaced, and of little use. Mining companies tended to be prospect focused, lacking a real understanding of the regional setting, relying on prospect geology and geochemistry with only a cursory use of geophysics. Exploration programs tended to be directed by senior management. Exploration of covered areas was mainly confined to chasing outcropping mineralised structures out into the pediment. Substantial areas of likely shallow cover had received little or no attention. There appeared to be a real opportunity for Geopeko to apply an integrated regional exploration approach, concentrating on covered areas, an approach that had been so successfully applied at Goonumbla and Lake Cowal.

A compilation of all available data (geology, magnetics, gravity, mineral occurrence, DEM, satellite imagery, interpreted depth of cover, interpreted structure etc.) began immediately and resulted in the compilation of a series of transparent overlays of standard maps of Nevada to form a manual GIS (GIS software was not available at that time). Integration and interpretation of these data in conjunction with numerous field trips resulted in a number of targets being developed, primarily in the pediment adjacent to, and/or along strike from, significant or developing gold mines. Several target areas were pegged or acquired through Joint Venture.

Reconnaissance geophysics and shallow reconnaissance drilling was undertaken to the north and along strike from the newly discovered Ruby Hill deposit at Eureka and on the pediment to the south of Round Mountain. The most exciting project was at Fire Creek, where a detailed HEM survey not only outlined the argillic alteration system around the high grade Fire Creek epithermal vein but indicated a number of new alteration systems. In addition, a regional HEM survey suggested that Fire Creek occurred on the same structure as the Mule Canyon deposit to the north. The regional survey also suggested that the pediment to the east was relatively shallow and that the structure that hosted the newly discovered Pipeline deposit, 20 km to the south, continued to the north and into the ground controlled by North. Although narrow zones of gold and areas of argillic alteration were intersected it was not sufficient to justify further work on the main prospect. Reflecting changing strategy, no drilling was undertaken on the pediment to the east.

Exploration in North America was well funded. Sub-bases were established in Reno and later in San Louise Potosi, Mexico, and local staff were retained. Both offices remained open until the Rio Tinto takeover. Significant work was undertaken but the focus progressively changed after 1994 to joint venturing into established projects, and to assisting the acquisitions group in assessing development projects in Canada, Mexico, and Central and South America. Funding for grass roots and early stage exploration projects became progressively harder to obtain.

NEW CALEDONIA

The first move into overseas exploration had been in New Caledonia. Participation in exploration for base metals and gold there was offered by Claude Dufor, a surveyor with a consulting business in Noumea. A visit to the area confirmed that the potential for mineralization typical of volcanogenic massive sulphide deposits, known well to Geopeko through the work of Large, Taube, and others, was present in terrain where airborne electromagnetic surveys should be able to detect them, so a proposal was put to PWL to farm into the Dufor properties.

Late in 1981 an airborne EM survey, using helicopter-mounted equipment, had been approved. It was substantially completed by March, 1982, and plans were being made for drilling of existing prospects in mid-year, and for follow up of anomalies from the survey. The airborne survey results, when they arrived, were not spectacular but were sufficiently interesting to warrant ground follow-up, but it wasn't to be. PWL terminated the project. Geopeko was told that the Board could not support the project further because off-shore exploration expenditure could not be used as an Australian tax reduction.

CHILE

In 1992 Geopeko had been assessing the exploration potential of a number of countries with the aim of establishing a second overseas base, when North's Business Development group, set up to assess mines and advanced projects worldwide, took it into Chile. In February Rob McDonald, an independent mining advisor, brought to its notice the possibility that Codelco would be placing some of its advanced Chilean copper deposits on the market. North was immediately interested. In October, McDonald led a North team which included Broomhead, Peter Lester (the new GM – Business Development), and geologist Colin Sinclair to Santiago to meet with Codelco management, investigate the Chilean commercial and legal systems, and look for possible acquisitions or partners among the local smaller companies. Geopeko had been asked to provide an appropriate senior geologist for the team and Sinclair, with his involvement in the assessment of the Parkes porphyries and his extensive international experience, was a natural choice. The team liked what it heard and saw. El Abra, a large extensively drilled porphyry copper deposit with a thick secondary blanket, located only a few kilometres north of Codelco's huge Chuquicamata mining complex, would come up for purchase by tender early in 1993 and North would have a good chance of being included on the list of companies invited to tender. Broomhead was anxious for North to keep a presence in Santiago through the January-February holiday season. Sinclair got the job and, with little to do other than keeping North's flag flying in government circles, it was decided to have a look at Chile's exploration potential and land availability. Thus Geopeko, this time operating as Inversiones North (Chile) Limitada, opened an exploration base in Santiago by default.

Rolf Forster had flown in to assist Sinclair and hold the fort while Sinclair returned to Parkes to make appropriate arrangements for the running of Geopeko's Eastern bases during his extended absence. Probably as a test of North's technical ability, Codelco came forward unexpectedly with the invitation to tender on a grass-roots porphyry project in the northern part of the country called

Mamiña. Forster gathered as much information as he could and passed this on to Australia. Mamiña lay between two producing mines, Cerro Colorada, a porphyry deposit being open pit mined, and Sagasca, an underground mine on a copper deposit that had formed through precipitation of copper within gravels beneath an ancient stream channel. It was considered that the copper in Sagasca could not have travelled as far as from Cerro Colorada. Hence there might be a large porphyry deposit within the Mamiña block. The daunting feature of the project was the 100-300m thickness of semi-consolidated conglomerate and sandstone covering the target basement rocks. This would present a challenge to electrical geophysics and to drilling and, with two 100m deep gorges crossing the area and part of it highly dissected, a challenge also to gaining appropriate access and to being able to undertake ground geophysical surveys. Experience was drawn on from Parkes, Perth, and Melbourne, and from remote sensing consultants, to put together a staged exploration program and cost estimates, and the proposal was submitted to Codelco in May 1993. In June Sinclair heard that the Mamiña bid was a good one but might not be the best. Somewhat to Geopeko's relief, it wasn't.

Data on El Abra became available in March and Sinclair had to work hard to keep North on the list of invitees. Pat Stephenson, by then having given up the Chief Mine Geologist's job with North and set up a consultancy, was called in to separate secondary ore from primary and to prepare a resource estimate. In May an MMD team, including Stephenson, visited El Abra and put together a plan to develop the bid. The resource was encouraging with 25% more oxide than expected.

Sinclair met with Ron Kitching, by then residing in Antofagasta and acting as front man for several Chilean syndicates, to assess two properties they had on offer. He liked them and farm-ins were negotiated over the next few months to prospects K1 and K7. K7 was quite near El Abra. North's El Abra bid was lodged and in October Codelco announced that it was bringing in the International Finance Corporation as an auditor on the El Abra bids. North joined an Australian mining and financial delegation to Santiago in September and, following a visit to Bolivia by Sinclair and Bruce Hooper, Broomhead and Lester visited Kori Kollo gold mine in November to discuss a possible joint venture with Battle Mountain. Negotiations did not progress very far.

Sinclair was preparing prospects K1 and K7 for drilling early in the new year. In December he got notification that the Canariaco property in Peru was available for purchase or staged option. It was a porphyry deposit with a resource of 400 million tons grading between 0.5 and 0.8% Cu. North's executive was alerted and plans put in place for an assessment and submission well before the January deadline for bids, but Broomhead called it off in early January for the reason that there was insufficient time for a risk assessment. A commissioned risk assessment had been available for several weeks but with the Christmas rush Business Development had failed to pass it on to the executive.

This illustrates North's lack of risk-taking, but a better example was Bajo de la Alumbrera in Argentina. In 1990 Paul Heithersay had recognised this as an important porphyry copper deposit with good gold credits available at a very low cost because the small Canadian company which had it, and had defined a large resource, was struggling to survive. Geopeko recommended a detailed assessment but didn't get support. Late in 1990, it came up again through Austrade and this time was recommended by Legge. Again there was no response.

In 1993 agents representing International Musto, the Canadian junior that held it, discussed with Sinclair the sale of it to North for

$9 million. Again a negative response, except from the Company Metallurgist, J Johnson, who informed the company that Alumbrera could be put in the middle of the Brazilian jungle and still make money! The lack of risk-taking by Melbourne office in 1993-4 was further emphasised when the Company declined at the last moment to put in bids for four large Peruvian copper porphyries, for which Fujimori and his government were requesting development proposals and which are now producing. North management stated that they were not ready for Peru.

Three years later, but with little further exploration, North followed MIM into the Alumbrera project, paying $225 million for a quarter share.

Delays in assessment of the El Abra bids by Codelco were relayed to Sinclair in February 1994 with the recommendations reported to be going to the Codelco Board in April. In June the Phelps Dodge offer of over $300 million for 51% equity was accepted. North's bid was something less than half this figure. Everyone seemed to agree at the time that Phelps Dodge was paying too much. There was talk some time later of a renegotiation to give Phelps Dodge a better financial position but it got its foot in the door with the high bid and El Abra was developed and is still mining, with the Phelps Dodge 51% holding (but from 2007 as part of Freeport McMoRan). North did a fine job of analysing what was known and building a robust model based on this but it could not even contemplate 'blue sky' based on the geological knowledge of the resource let alone a little extra to win the bid.

Several prospects were drilled in the ensuing years with limited success, due mainly to large tracts of the prospective ground being held by the Chilean government.

NEW ZEALAND

Taking a lead from the group's interest in New Zealand's Westport ilmenite sands, Paul McInnes had applied for exploration licences in a number of coastal areas in the North Island of New Zealand. The Government was supportive but there was an adverse public reaction on environmental grounds. When the Government announced, in September 1991, a change in land title laws that would almost certainly allow conversion of the areas under application into Maori Land the follow-up to the grant of the applications looked much too hard and they were withdrawn.

Applications had also been lodged over a block of Cambrian volcanics in north-west Nelson, as possible equivalents of the highly mineralized Mt Read Volcanics of north-west Tasmania. They were allowed to stand and eventually were granted and, following an extensive helicopter EM survey, explored on a campaign basis from Australia. Early results were not encouraging and, faced with of increasing environmentalist resistance and a threatened national park over the whole area, Geopeko withdrew from the project.

MALAYSIA & INDONESIA

At much the same time as the NZ ilmenite activity in the early nineties, Imants Kavaleras, the expert at the time in SW Pacific super-porphyries and well known to Heithersay, put in his report on assessment of Geopeko's Malaysia proposal. He wasn't enthusiastic about the potential of Sarawak and recommended a focus instead on Kalimantan on the other side of Borneo. Considering that Indonesia would be a much more difficult sell at Board level Geopeko management decided to defer a second move into overseas exploration until there was more geological information and a better risk assessment for Indonesia and other countries of interest.

In May 1993 Newmont invited North to consider farming into its Batu Hijau porphyry copper-gold project in Indonesia. North took the offer seriously but in the end did not make an offer. This did, however, indicate to Geopeko management that there might be an easing in attitude towards operating in Indonesia and it was decided to develop a new proposal for entry into porphyry exploration in that country. To gain a better appreciation of country risk, but also to have external support when submissions were made to the Board, Geopeko had worked with the Business Development group to commission an independent risk analysis. With Indonesia, it went much further, taking senior commercial and legal staff from North to Jakarta on a fact-finding trip before developing the submission. It was in its final draft form in January 1994 but had to be put aside when Williams was retrenched. Later in 1994 Indonesian exploration was raised again, gained Board approval, and Heithersay was given the commission to open a base in Jakarta.

EUROPE and AFRICA

During 1995 North Limited were actively seeking opportunities for overseas expansion. Apart from iron, copper and gold, zinc was also a target commodity for the company. It developed an interest in the Zinkgruvan mine in Southern Sweden and a due diligence team was assembled to look at the possibility of acquiring this operation. Wayne O'Neill, as an exploration geologist, was part of this team and his role was primarily to provide an assessment of the potential for additional resources within the immediate mining area. O'Neill concluded that there was potential for significant additional resources and that exploration within the near mine environment was incomplete. These conclusions fed into the overall assessment of the operation, which was acquired by North

Limited in late 1995. Significant additional resources were identified in the mine, sufficient to replace the mined tonnage for many years. An exploration office was established near the mine and was headed up by Australian geologist Leigh Schmidt. It operated for a limited time but was not successful in delineating new resources away from the mine.

With North Limited still keen to "internationalise" O'Neill was given the position of Manager Exploration Europe/Africa in December 1995. The company entered a joint venture in Burkina Faso West Africa with South African based company Randgold Resources. Structurally controlled greenstone-hosted gold mineralisation was the target. The JV operated from existing Randgold offices in the capital Ouagadougou. North supplied additional geological, geochemical, geophysical and structural expertise. A number of major gold anomalies were discovered and tested but failed to make the grade requirements and the 2Moz target established for the area. A gold price languishing in the mid US$300 range was not encouraging. Two new grass roots discoveries identified by the JV are now being mined by Russia-based Nordgold. They are the Bissa Mine opened in 2013, and the Bouly Mine opened in 2015.

North registered an exploration company and established an exploration office in Abidjan, Cote d'Ivoire, and became involved in several joint venture opportunities. At the time of Rio Tinto's takeover of North it had just received some very encouraging results from the first round of diamond drilling completed on the Tongon deposit in northern Cote d'Ivoire. This was another JV with Randgold Resources. Randgold went on to put the Tongon Mine into production in April 2010 and at the end of 2015 stated proven and probable reserves stood at 2 Moz.

In September 1998 O'Neill established an exploration office

in Pangbourne in the UK to service the exploration activities and project assessments being conducted throughout Europe and Africa. These activities were carried out in Burkina Faso, Mali, Cote d'Ivoire, and Senegal in West Africa for greenstone-hosted gold. Zinc opportunities were pursued in Ireland and Greece and other opportunities were pursued in Sweden, the Slovak Republic, Romania, Hungary, Spain, and Italy. At the time of the Rio Tinto takeover a reconnaissance trip had been taken to Iran and relationships were being developed with various contacts. Following the takeover in October 2000 all of North's assets were handed over to Rio Tinto representatives and most North employees were repatriated to their home countries.

CHRONOLOGY OF EVENTS

Year	Event	Orebody Found	Orebody Required	Comment
1949	Peko (Tennant Creek) Gold Mines No Liability floated		Peko	
1954	Name change to Peko Mines No Liability			
1955		Black Angel		Later mined by Normandy
1957		Orlando		
1961	Peko-Wallsend merger			
1962	Exploration extends to Kimberley	Ivanhoe Warrego	Mt Angelo	Mt Angelo shown to Ryan by lessee
1963	Geopeko Limited formed. Tennant Creek base set up. Prko Mine services set up.		Little Mt Isa	Found by BMR - right place at the right time?
1964	Tanami-Granites exploration begins	Juno		Orlando Mine opened
1965	Exploration expands to Top End of NT, Arunta, NW Queensland, NSW		Strathorpe-Inverell	Ivanhoe Mine opened
1966	Darwin base opened. Tomago Sands Project begins			
1967				First Seminar at Tennant Creek
1968	Mt Morgan takeover, base opened. King Island Scheelite takeover, base opened.		Mt Morgan Bold Head Mt Bundey	Juno Mine opened
1969		Ranger 1 Gecko	Woodcutters	Woodcutters later mined by Nicron Resources
1970	Jabiru base opened			
1971	Coal team established	Dolphin		Death of Lew Richardson
1972	Chatswood office opened			
1973	Parkes base opened. COal team closed	Argo		Warrego Mine opened

Year	Event	Orebody Found	Orebody Required	Comment
1974	Perth base opened. Townsville bases opened.	Rover 1 Endeavour 7		Sir John Proud retires as CEO. Austirex contract in Iran begins. Dolphin Mine opened.
1975	Chatswood office closed, move to Gordon		Mt Chalmers	Federal Government acquires an interest on Ranger One
1976		Ranger 68 Ranger 10		
1977	Devonport base opened. Tennant Creek Mine Services closed.	Northparkes Abra		
1978		Surveyor One	Mt Bonnie	Sir John Pound retires. Surveyor One later mined by Kagara Zinc Limited.
1980		Peak Hill, WA		Mt Chalmers Mine opened Gecko Mine opened
1981				Ranger One mine begins production
1982			Robe River	
1983	Brisbane base opened. Mt Morgan Base closed	Rosehall		
1984	King Island base closed. Devonport base closed. Darwin base closed.		Woolwonga	Elliston leaves Peko-Wallsend
1985		Kanowna QED		
1986				Argo Mine opened
1988	Peko-Wallsend taken over by North Limited	Lake Cowal Colmer		Peak Hill (WA) mine opened

Year	Event	Orebody Found	Orebody Required	Comment
1989	Cobar base acquired from Norgold	Kanowna Belle Mayday North McKinnons Tank		Kanowna QED (Golden Valley) Mine opened
1990	Head Office moved to Melbourne. Ballarat base opened. Adelaide base opened. Santiago base opened.			
1992	Tennant Creek base closed. Ballarat base moved to Ararat. Mt Isa base opened. Darwin base reopened. Denver base opened.	Endeavour 48 (became part of Northparkes)		
1993	Gordon office closed. Adelaide base closed.			Kanowna Belle Mine opened.
1994				Brian Williams retrenched. Northparkes Mine opened.
1995	Ararat base closed. Lima base opened. Jakarta base opened. Mt Isa base closed, transferred to Brisbane.			Geopeko becomes North Exploration
1996	Brisbane base closed. Zincgruvan, Swedan base opened.			
1997	Abidjan, Ivory Coast base opened.			
1998	Brisbane base reopened. Pangbourne, UK base opened.	Bissa Bouly, Burkino Faso		
2000	North taken over by Rio Tinto. Brisbane base closed	Tongon, Ivory Coast		
2001	Melbourne office closed. Parkes base closed. Perth base closed.			

INNOVATION AND INVENTION

It is a truism in the mining industry that junior exploration companies, who are commonly run by earth scientists and other mining professionals, have a better discovery rate per exploration dollar than the majors, whose directors are more likely to be drawn from the legal, accounting, and business professions and have less of the background and experience needed to get the 'feel' of a proposed exploration project and the likelihood of a successful discovery.

Geopeko's track record of discovery speaks for itself. That it went almost unnoticed by its competitors is interesting but is not relevant to the success. That can be attributed to the culture of excellence fostered by Elliston and Richardson and wholeheartedly supported by Proud. Had the early board of Peko-Wallsend not given the Geopeko team its head, at times in the face of quite serious opposition at board and joint venture level, the story could have been quite different.

The culture extended beyond the science. From orienting drill core in a tray to preparing a drill hole survey, to collecting a sample, the utmost professional rigour was demanded: and as the technical employees saw that it worked so they embraced the culture, so that the field technicians, the drilling supervisors, the drafting staff, the laboratory hands, the library assistants, became part of the professional team and took pride in what they did. Of equal importance was Proud's policy that it is the geologist's job to find the orebodies, not the Board's, and that they should be left to get on with it and their judgement backed.

There was also a strong culture of innovation within the company. The early success at Tennant Creek was largely due to the innovative approach to modelling of the magnetite orebodies that Lew Richardson had developed.

But the use of magnetometry and other geophysical methods went well beyond that of drill target generation. Refined by Bob Richardson and his team they became geological mapping tools whereby the project geologist, the geophysicist, and the geochemist jointly took the time needed to compile all relevant geological information from the emerging data and assess its significance. That was vital in areas with substantial alluvial and/or sedimentary rock cover, as well as in areas with reasonably mapped outcrop, and provided an enhanced understanding of the geology, more reliable interpretation, and hence the selection of regions of greater prospectivity and better targets for drill testing.

An important next step was essential – that of calibrating the interpretation with drilling or other ground observations. The process of extracting as much data as possible from the surveys was important, not only in geological mapping and target generation but in team building, and could only happen when the scientific and technical staff were allowed sufficient time and opportunity to conduct proper analysis of the data and to fully utilise the information obtained. The evaluation included a policy of additional drill holes until the team was satisfied that it had enough data to arrive at a proper assessment of any project.

The process was initiated in the Tennant Creek area, was somewhat refined for the Rover field next door, and by the late 1970s was widely used by the exploration teams across Australia.

It was the foundation of the highly successful campaigns on the plains of the Alligator Rivers Uranium Province where outcrop was virtually absent, and in NSW where it led to the discovery of the Northparkes and Lake Cowal deposits.

Geopeko was among the first Australian companies to develop geostatistical resource estimation on its orebodies, to use satellite imagery, to embrace new methods of sample analysis, and to build and deploy a Transient Electromagnetic survey ("TEM") system.

Austirex Aerial Surveys, when it was formed as a PWL subsidiary, led the world in aerial radiometry and data processing techniques, an attribute that led to the successful bid for the contract in Iran[23].

A vehicle-borne navigating system was developed before GPS became available and made a significant contribution to the ability to locate and measure anomalies quickly in remote areas where there was no outcrop, giving the company a competitive edge when new data were released by government geological surveys. The vehicle also allowed rapid accurately-positioned reconnaissance surveys of localities of interest, a major improvement from the earlier techniques of hand cutting lines for theodolite or compass gridding.

The drills that became the highly successful UDR rigs, now to be found world-wide, were designed by Ron Kitching and were first built by a PWL subsidiary; and the 'auger-core' drilling technique for mapping below thick cover was developed for use on the plains of the Alligator Rivers Uranium Province and subsequently used by other bases to carry out subsurface mapping.

And as the more junior members of the company saw that innovation was not only accepted, but was encouraged, so they understood that they could introduce new ideas without fear.

---

[23] Appendix 2.

ORE GENESIS

Successful exploration is underpinned by a through understanding of the ore-forming process in the target province: whether it be tropical leaching that produces laterite, and hence bauxite; processes deep in the earth that generate ore-forming fluids that ooze up into the crust and create orebodies; or other processes by which sought-after elements are concentrated to economic levels. A rigorous examination of the rocks available for inspection is fundamental to understanding what processes may have taken place in any mineral field. Without the geological input the analysis of geophysical and geochemical data may be imperfect and, equally, the application of data from the other disciplines enhances the understanding of the target geological environment.

This process underpinned the success on the Tennant Creek field. Doubts as to the origin of the porphyroids there had first been raised by BMR field geologists, who had completed the first regional mapping program of the field in the early nineteen fifties. Team leader John Ivanac also suspected that these crystalline rocks might have something to do with the mineralisation on the field. As Peko geologists extended their geological knowledge around the field those doubts intensified. Two main lines of evidence were: the absence of any evidence of thermal contact effects that should have been obvious if molten magma had invaded the country rocks; and, in many outcrops, a gradual transition from porphyroid to un-metamorphosed country rock, with evidence of the intermixing of unaltered sedimentary rock with the intrusive rocks. As work proceeded the conviction grew that these rocks were meta-sediments, but, as was the case with Wegener's early work on Continental Drift, a mechanism was lacking.

With the backing of Proud and Professor S Warren Carey

from the University of Tasmania, one never to shirk a scientific conundrum, Elliston set out to find a mechanism. In time he was introduced to Professor A. E. Alexander at the University of Sydney and Professor T. W. Healy at the University of Melbourne, and to colloids. As the trio collaborated on the subject it became clear that all fine-grained sediments pass through a colloidal phase as they dehydrate, and during that phase they can, if subject to appropriate processes, recrystallise into 'igneous-looking' rocks. During that process the contained mineral elements that do not fit into the newly crystallised minerals are expelled, together with the water in the sedimentary pile, which is heated up, as the process is exothermic, thus generating 'hydrothermal mineralisation'.

Elliston had found his mechanism, not only for the formation of the igneous-looking rocks but also for the origin of the mineralising fluids at Tennant Creek, and, by implication, anywhere else in the earth's crust, and has devoted his ensuing years to the study of this process. He has published widely, at times in the face of fierce, and very unscientific, opposition from the established geological fraternity. Any doubts he might have had at the beginning were put to rest, however, by the visit of Professor Tom Barth from Norway, at that time the world's leading expert on granite genesis, who visited Tennant Creek and wholeheartedly endorsed Elliston's work.

The ability to recognise that ore-forming processes had taken place meant that the ore-generating potential of any suite of rocks could be assessed. In particular, Elliston's leadership and methods taught the company's scientific personnel to look closely at every rock and to try to understand what its history was. This approach lay behind much of the successful selection, by Geopeko geologists, of suitable mineralised provinces as exploration targets.

THE RICHARDSON CONTRIBUTION

Lew Richardson was a pioneer of exploration geophysics in Australia. He died suddenly in 1971, but by then he had created a specialist group that grew to be a real force in the science of applied geophysics in Australia and made an important contribution to Geopeko's exploration success.

One of his most important projects, as a member of the original AGGSNA team in 1936, was the completion of extensive magnetometer surveys over areas of known surface mineralisation on the Tennant Creek field, including the Eldorado, Great Eastern, Rising Sun, Peko, Golden Forty, Mt Samuel, Orlando, Black Angel, and Wheal Dorea areas.

The Tennant Creek orebodies are not easy drilling targets:

- many are in soil-covered areas with no outcrop to provide geological guidance;
- the causative bodies are relatively small (but high grade), of irregular shape, and easily missed by drilling, particularly as holes on the Tennant Creek field tend to deviate from their planned path due to the texture of the host rocks;
- most of the bodies are steeply dipping and "pipe-like", lie at depth, and require accurate inclined drilling for a reliable test;
- the high magnetic susceptibility of the magnetite lodes causes a phenomenon called demagnetisation. The original modelling algorithms could not take demagnetisation into account and gave erroneous interpretations.

It soon became evident to Richardson that the available interpretative models were inadequate. At the time, interpretation methods were relatively crude. Magnetic anomalies could be modelled by comparing observed anomalies with anomalies due to theoretical magnetic bodies shaped as spheres or simple tabular shapes. Together with University of New South Wales mathematician Bruce Kirkpatrick Richardson developed algorithms and routines that enabled the modelling of magnetic anomalies caused by ellipsoidal shaped bodies. It was a rigorous algebraic solution that took demagnetisation into account.

As an example, his analysis of the Peko No 1 anomaly showed potential for magnetic ore at depth below the oxidized gold-bearing body and he recommended and targeted the drilling that discovered the rich copper lode that set Peko Mines NL on its path to success.

At the conclusion of the World War 2 he formed his own geophysical consulting group, L A Richardson & Associates ('LAR'), and in the ensuing years his group provided consulting services to many of Australia's mining companies. Under his son, Bob Richardson, the group grew rapidly through the 1970's to a team of more than 20 people including geophysicists, physicists, mathematicians, electrical engineers, surveyors, data processors, and field technicians. The Richardsons had concluded that many earth science problems are too challenging for one person to solve – more is achieved by having a small group of complementary disciplines driving the same pencil.

During this period the capability of LAR expanded to embrace all geophysical methods including magnetometry, gravimetry, resistivity/Induced Polarisation, Electromagnetic and Self Potential techniques, down-hole methods, radiometry, measurement of the physical properties of drill-core, and data processing. This growth paralleled Geopeko's expansion into other regions of Australia where it was exploring for a wide variety of commodities and deposit types in a range of geological environments.

In 1978 LAR was acquired by Peko–Wallsend and morphed into the Peko Geoscience Division. Until that time it had reported separately to Peko-Wallsend management. That was a wise decision by Proud to ensure that the geophysicists functioned alongside the geologists and not beneath them. Because the geophysicists had a long-term involvement in Geopeko's exploration programs they had the benefit of following programs from concept, through regional surveys, to prospect testing and mining. It was a learning experience that made them better geophysicists. Many other companies use geophysicists, and also consultants in other fields such as geochemistry, as "specialist" consultants whose expertise is called on at irregular intervals when the geologists think they need it. Richardson and Proud believed that they should operate as an integral part of the modern exploration process, involved in conceptual thinking, exploration planning, data evaluation, management, and feedback of results, working hand-in-hand with geologists on an ongoing basis. This arrangement was of immense value to the geologists as well.

THE INSTRUMENT MAKERS

The LAR group developed and made their own instruments when they could not find any that met their requirements. In addition to those listed below, they introduced a scintillometer-based assay device at Ranger One and developed an astatic drill-core magnetometer.

Aerial Geophysical Surveys

The Group provided the management and most of the technical expertise that led to the development of the Austirex airborne systems and their use in Iran for what was, at the time, the world's largest and most advanced airborne geophysical survey and data management system.[24]

---

[24] Appendix 2.

The Chobham Navigator.

As exploration ranged beyond Tennant Creek to more remote areas, where available maps were poor, road s rare, and the terrain flat and featureless, such as Tanami-Granites and the Rover field, navigation to locate anomalies that had been revealed by aerial survey was very difficult. The standard technique had been to use compass and odometer to navigate through the bush during the day, followed by star observations at night using a theodolite to get a location fix. This was a slow and cumbersome approach and not at all accurate. By the time the teams had veered around several mulga forests, claypans, and creeks they might be kilometres away from their planned destination.

Astronavigation was a frustrating process. It was necessary to find a suitable star, get four accurate fixes on it, while holding a torch to illuminate the theodolite scale in one hand, field book and pencil in the other, and twiddling the theodolite knobs, and then sit around the campfire doing the reductions using the star almanac and seven figure log tables, under the feeble light of a gas lantern. Furthermore at best it would only give an accuracy of about 500 metres, so then it was necessary to walk the area back and forth with the magnetometer to find and outline the anomaly.

The team soon got tired of this approach and after some digging around in the literature discovered a brilliant device called a Chobham Navigator, manufactured by Sperry Instruments in the UK. It had been developed for the British army and one was being used on the Woomera rocket range to locate and recover spent rocket casings. The system used two electronic fluxgate compasses on a boom on the rear of a Landrover to obtain heading information from the earth's magnetic field. This heading information was combined with information from the odometer cable and interfaced to a mechanical computer called a "ball resolver" that calculated east and north displacement. By setting the known starting coordinates of the vehicle the device would then indicate a position on mechanical readouts in AMG coordinates with an accuracy of about 0.25% of the distance and direction travelled. It was a beautiful piece of equipment, built to milspec standards, and it revolutionised the navigation in remote locations before the advent of GPS. Two vehicles were fitted with these navigators, and were christened 'Blossom'.

Vehicular Magnetometry

The most common purpose for navigating to remote areas was to locate, measure, and assess specific magnetic anomalies. So it was a logical next step to equip the Chobham-navigated vehicle with a magnetometer. The obvious technical choice was the proton precession instrument that measured the total field so that the sensor did not require orientation – essential for a vehicle bumping along in the spinifex. The principle of proton precession, and its use for magnetic measurement, was invented by Varian Corp in the early 1950's, but there was no suitable instrument for Richardson's purposes on the market. Mike Palmer, Richardson's instrument maker, retired to his workshop in Walcha, NSW, and designed and built a cycling proton precession unit, packed it up in a box and sent it to the Sydney office by rail. Disaster! – it never arrived and was never seen again. Following extensive enquiries with the railways, and meetings with insurance agents, Richardsons gave up and started again. The second unit was completed and fitted into the Navigator vehicle, with the sensor mounted on a boom projecting from the rear of the vehicle and extended to minimise the vehicle's magnetic effects. The aluminium body of the Landrover helped.

The system, as mounted in the Landrover, consisted of the Chobham Navigator (between seats), magnetometer, chart recorders, and various control units. Noise from the engine electrics sometimes swamped the precession signal and that was overcome by building an astatic sensor, ie. a sensor with two coils so that electrical noise was cancelled out. The magnetic effect of the vehicle was reduced or at least modified to a simple sine curve relationship between vehicle heading and the magnetic effect at the sensor, by fitting a large de-gaussing coil carrying DC current in the tailgate of the vehicle. Magnetic field strength was recorded on a chart recorder that was slaved to the motion of the vehicle. The vehicle acquired the name of 'Blossom'.

Richardson's were the first, and possibly the only, outfit to build and use such a system in Australia.

Vehicular Radiometry

Following the Ranger discoveries a mobile radiometric system was set up in one of the navigator vehicles. It comprised a 3x3 inch sodium iodide detector coupled to a Bertholdt scintillometer and chart recorder. This system saw extensive service on the featureless Munmalary plains where it was employed to lay out reconnaissance gridlines and to locate radiometric features that were evident on the aerial maps. In areas of no outcrop and thick vegetation it was of tremendous assistance to the field crews and substantially less costly than line clearing and theodolite surveys would have been.

Down-hole Magnetometer

For many years the need for a down-hole magnetometer to provide another dimension to magnetic measurements, provide information that would assist in refining the modelling of magnetic bodies that had been drilled, and help find a body if it had been missed by the drilling, had been recognised. They had been used in the oil industry for many years, but only in large diameter, and usually in near-vertical drill holes. A system for use in the smaller, often inclined, holes was more of a challenge. Bob Richardson had visited Develco in the US and was aware that they could supply relatively small fluxgate sensors. The Richardson team built a working system that saw very useful service at Tennant Creek and elsewhere.

The instrument had considerable use in the Central Field over the next few years. It was routinely used in the drilling of deeper ironstones to gain an immediate indication of geometry and size. A number of old holes were reopened to get a better magnetic profile. The discovery of ironstone with ore-grade gold at considerable depth just to the east of the White Devil mine was directly attributable to its use.

Portable Transient Electromagnetic (TEM) System

Richardsons were the first to design and build a portable TEM system in Australia. The only other system in use in Australia at the time was the Russian MPP01, imported and used by Western Mining Corporation. The geophysicists managed to get a good look at one of these and met the Russian inventor Mr Bulkakov on his visit to Australia. Through his interpreter they learned a great deal about the system. Mike Palmer was less than impressed with the outdated Russian technology, which used germanium transistors, was convinced that a more sensitive and more portable device using modern silicon transistors could be built, and was successful.

GEOSTATISTICS

Efforts to improve the quality of reserve and resource estimation had begun in Tennant Creek in the sixties, triggered mainly by Board disbelief that the resource estimates being reported from the extraordinarily high grade Juno orebodies were correct. Dr Sichel was recruited from South Africa to carry out a technical review, went to Tennant Creek, and reported to the Board that the work being done there was reliable. Perhaps the fact that this work had improved the reconciliation with production results also impressed the Board.

When Ranger One was discovered it was immediately apparent that here was an orebody that demanded the estimation of the resource using the new methods to complement the traditional polygon-based methods. Proud's stated objective in confirming the appointment of Terry Quinlan to head up a resource estimation unit was that "There will be no question regarding the reliability of the Ranger One ore reserve estimate", a statement brought on by the fiasco at the Nabarlek uranium deposit nearby where the original published reserve estimate had been written down by about three quarters. Elliston was also concerned with the Ranger One estimates.

At that time the development of geostatistical methods of ore estimation was led by Michel David and Andre Journel, and Quinlan was despatched to join in a meeting between Geopeko and Renison Goldfields to learn as much as he could. David had been brought to Australia by the Australian Mining Industry

Research Association. Geopeko geologists from various bases then joined Quinlan at a short course presented by David at the Australian Mineral Foundation in Adelaide. Quinlan recalls "that these methods appeared daunting and there was a lack of detail concerning the implementation of the methods". He set out to develop a computer program that would accomplish it.

The computing technology was not as user friendly as is the case today where all staff have access to, and can manipulate, the data as necessary. The process of introduction to existing operations involved significant changes and was difficult, not least because it added substantially to individual workloads, as it was being developed in parallel with existing systems. Teething problems had to be managed and it is testament to the patience of all concerned that progress was made – albeit slowly at times. However the goal of being able to model orebodies to provide reliable tonnage and grade predictions for ore blocks that matched the production schedule was "the light on the hill" for all concerned.

Quinlan's earlier work, as a contractor prior to his engagement by Geopeko, had involved an 'Inverse Square of Distance' method to interpolate, from drill hole data, grades at points on a three-dimensional mine grid, followed by numerical integration to obtain a grade estimate for a block of ore. He had developed this approach for the Mount Cannindah orebody, working with Mt Isa Mines. The method was to be compared with estimates prepared by Mt Isa geologists using manual methods, and by BRGM using geostatistical methods. This exercise had given him valuable insights into the comparative utility of the various methods. The Ranger One No 1 Orebody was also estimated initially by both methods.

Quinlan then set out to develop the necessary program, which was later named 'TORRES'[25], and to apply it initially to the Ranger

---

[25] The Ore Resource and Reserve Estimation System.

One estimates. That estimate supported PWL's application to Schroeder Wragg for development finance. They in turn retained Fluor, who commissioned Andre Journel to assess the Geopeko estimate. Journel gave it the seal of approval, thus beginning a long and satisfactory association with him.

Later, when the No 1 Orebody at Ranger One was exhausted, a reconciliation demonstrated that Quinlan's estimate had been within a few percent of the final product.

RADON GAS AS AN EXPLORATION TOOL

The discovery of the uranium deposits in the Alligator River Province encouraged Geoff Sherrington, a geochemist with a stake in a geochemical laboratory in Queensland, to visit the Australian Atomic Energy Commission at Lucas Heights to learn more about uranium exploration. His interest led to his recruitment into Geopeko to head up a geochemical division.

During one conversation with scientists at AAEC it was pointed out to him that one of the uranium decay products was radon-222, a radioactive gas, which releases alpha particles as it decays. The theory was that radon, being an inert gas, might escape from a concealed uranium deposit located far beneath the ground, travel upwards along suitable pathways, and be detectable remotely as an exploration guide. The scientific literature started carrying articles about how marvellous radon was as an exploration tool. It was even claimed that one could find oil deposits under the sea using it. It was also claimed that uranium deposits lying at hundreds of metres depth on the Colorado Plateau had been located by radon survey, a claim viewed with some scepticism by many in the field.

The AAEC people told Sherrington of a French Kodak product, a red film on a clear plastic base, named LR-115. When alpha particles hit the red layer of cellulose nitrate (the stuff of old film stock) they damage the polymer and leave it weakened, and if the polymer is immersed in hot caustic soda solution over an appropriate period the damaged film is dissolved to leave a white spot on the red background which can be viewed, and counted, under a microscope. The more spots in a given area, the higher the radon concentration for a fixed exposure time. Sherrington developed a radon detection device based on this information and called it the "Alphalogger". The idea was to attach a short length of film to the inside of the base of a paper cup, and bury the cups upside down in the soil to a shallow depth to intercept any radon emanating from a concealed deposit of uranium that, due to the depth of soil and rock cover, could not be detected by scintillometer.

At Ranger One a trial line was run over the Number 3 ore body, which was exposed over an area of many hundred square meters. There was a positive result from the Alphaloggers which was reported in a paper that Sherrington delivered in Canada in May, 1982 and subsequently published in the *Journal of Geochemical Exploration*.

Another company, Alpha Nuclear, had developed an electronic alpha particle detector. Some Alpha Nuclear meters were acquired and tested but the results were unsatisfactory as it was found that the levels of radon fluctuated wildly throughout the day due to the influence of wind and temperature changes. It was evident that only measurement over an extended period of time would work.

At the time the Westinghouse Corporation was offering a service known as 'Track-Etch', which was based on the same principle as the Alphalogger. Geopeko had looked at the service. It was horrendously expensive and Sherrington was of the opinion that the Westinghouse patent would not stand up to examination in the patents court if the Geopeko method was challenged. It was classed as "prior art" in the language of the patent attorneys. PWL backed the development of the Alphalogger, a laboratory was set up at Jabiru, and detectors were set out on the alluvial plains of the Munmalary license areas. Positive results were obtained but they tended to reveal the locations of geological features up which the gas could travel rather than the exact position of the source of the radiation below. Nevertheless, the follow-up of positive results led to the discovery of the Ranger 68 uranium deposit under some 30m of Cretaceous cover.

Lead isotopes as a guide to ore

Geopeko had long used an empirical approach to assessing prospectivity of ironstones at Tennant Creek, based on the mineralogy of the lode types. Magnetite-chlorite lode was prospective for gold-bismuth; magnetite-quartz was prospective for copper; talc-magnetite lode was not prospective at all. But such crude measures were not adequate for what was now needed; a method of ranking an ironstone's prospectivity from the sampling of core from a single or a limited number of drill holes. Building on a BMR survey which had reported multi-element analyses on samples from most outcropping and some subsurface ironstones, these and other multi-element data were statistically analysed, but offered no clear path to discrimination.

Assistance was sought from Ross Large, by then holding a Chair in Geology at the University of Tasmania and several years into developing the Centre for Ore Deposit and Exploration Studies (CODES). He suggested the study of isotopes across a range of mineralised to unmineralised lodes, and additional trace element studies. Sulphur and oxygen isotope determinations were to be done at CODES. For lead isotopes, he recommended an approach to Brian Gulson of the CSIRO in Sydney who was embarking on research into the possible fingerprinting of various types of mineralization using ^{208}Pb/^{204}Pb and ^{206}Pb/^{204}Pb ratios.

Geopeko sponsored the CODES studies on the suite of samples already held by Large from his PhD study, broadened to include more country rocks and ironstone from prospects that were thought to be non-prospective. The first batch of samples for a sponsored lead isotope study was collected in the latter part of 1984, from drill core and outcrop, to represent the known ore deposits (Peko, Juno, Orlando, Argo, and Warrego), the still hopeful prospects, and the ironstones written off as barren: twenty eight locations in all.

Gulson's team was the first to report results. Plots of ratios ^{208}Pb/^{204}Pb vs ^{206}Pb/^{204}Pb for all samples from the mines Peko, Juno, Orlando, and Argo (and one of the older small gold mines, The Mount) lay within a very restricted field, named by Gulson the 'High Pb Target'. Ratios from Warrego extended out from this field at similar ^{208}Pb/^{204}Pb values to much higher ^{206}Pb/^{204}Pb levels, defining a larger field termed the 'High U/Pb Target' and attributed to an overprint of mineralisation from the nearby granite.

Of the seventeen prospects ranked Priority 1 on lead isotope results, only five, Black Angel, West Peko, and Eldorado Anomalies 1, 3, and 5 had ironstone bodies of sufficient size to have hosted a gold orebody of significant size. The four still in Geopeko's hands were drilled further and three recorded some, but not sufficient, ore grade mineralization. Eldorado Anomaly 1 did not but it could be argued that the closely associated gold lode of the exhausted Eldorado Mine was part of the Anomaly mineralisation. The Carraman, a deposit with a positive return, was not available, but was later drilled to depth and mined by Giants Reef Mining.

Isotope studies by CODES offered little help in discriminating ore-bearing from barren bodies but did provide information in relation to ore fluids and ore emplacement. Trace elements did offer some potential. Large suggested that lead content alone might be as good a guide as lead isotopes and, in 1987, argued that ironstones with Pb > 20ppm and Zinc ratios (100Zn/(Zn + Pb)) in the range 0 to 10 had a good chance of hosting gold mineralization.

With the recruiting of a number of sponsors (which included CRA, BHP Minerals, Placer, and WMC) the CODES research program morphed into an AMIRA project looking at iron oxide copper-gold deposits at both Tennant Creek and in the Starra district of NW Queensland, and was a forerunner to research into IOCG mineralization in general. As part of this project, and armed with new more precise gold determinations, Stoltz, in 1991, revisited Large and Robinsons' trace element work, set new background upper limits for Au, Cu, Bi, Mo, and Pb, and concluded that "the likelihood that a particular ironstone contains a mineralized pod is significantly increased if analysed samples have anomalous concentrations (significantly above background) of one or more of Cu, Au, Bi, Mo, or Pb". That was about the time that Geopeko's involvement in Tennant Creek exploration ended.

In Western Australia a similar approach was used to develop chemical signatures and to classify the various types of volcanic rocks in the Archaean. Terry Ballinger, Ray Twist, and Bob Perring investigated the chemical difference between tholeites, komatiites, and high magnesia basalts, using the results to classify the various sequences and to identify those with potential for the presence of proximal and distal VMS deposits.

OPTIMISING GEOCHEMICAL DATA MANAGEMENT

A team led by Terry Ballinger developed a highly sophisticated data storage and retrieval system for the analysis of geochemical data, labelled GeoDBase, before Oracle, MS SQL Server, and other applications were available. Terry was very innovative in the storage and in assuring the quality of data so that he could make valid comparisons and provide valid analysis and interpretation. His work on Quality Assurance and Quality Control was designed to ensure that the labs were providing accurate and consistent values and that he had the data to prove it. He was a pioneer of low-level gold geochemistry and worked with the labs to provide meaningful data to be able to use this information effectively in exploration, especially for gold. This work led to the discovery of the Fiveways deposit and made a contribution to the discovery of other gold deposits in the Eastern Goldfields.

Terry's work on databases did not stop at GeoDBase. He contributed enormously to the revised geological interpretation, geological modeling, and resource estimation of the Mesa J CID Deposit at Pannawonica in the Pilbara. He set up a database in MS Access with full validation, internal integrity, and fully relational output. He set up the Access database for the West Angelas bedded iron ore deposit, which led to the development of a 400Mt+ high-grade iron ore deposit, which is one of the 'jewel-in-the-crown' deposits for Rio Tinto still to this day.

DECREPITOMETRY

In 1980 Kingsley Burlinson, a geochemist with Geopeko, developed a home-made decrepitometer, a small chamber in which rock particles were heated according to a pre-set rate and program, then cooled again. The basis of this technique is that when fluid inclusions are heated to the point at which the internal fluids homogenize to a single liquid phase, further heating will generate sufficient internal pressure in the inclusions to fracture the host mineral and cause a 'micro explosion', which can be detected by a pressure-sensitive instrument. Microphones on the small chamber record minute percussions as the rocks that had liquid/gas filled bubbles fracture when the temperature raised the pressure. A histogram of bursts versus temperature, as the sample is gradually heated, reveals information on the fluid temperature from which the host mineral phase formed.

This method is less precise than the microscopic observation of fluid inclusions, but a critical difference is that decrepitation does not require a transparent mineral and can be applied to opaque minerals such as magnetite and hematite. The decrepitometry patterns proved to be excellent indicators. A study of Tennant Creek samples demonstrated that there was a wide range of temperatures in the minerals at various mines, including oxidised samples that, to that time, had been considered to be due to oxidation during weathering.

To his credit, Kingsley did most of this work in his own time with very little financial support due to considerable (and unusual) scepticism within the ranks of Geopeko, and parted company with Geopeko as a result. He pursued his work independently thereafter.

GEOMETALLURGY

Geopeko was one of the first to introduce the practise of matching grade estimates and predicted metallurgical performance to the various rock types within an orebody, a discipline now gaining wide acceptance as 'Geometallurgy'. This approach recognises that the response of a body of rock to the mineralising processes, to the comminution processes, and to the processes whereby the valued components are separated from the gangue, can be affected by the geological and mineralogical characteristics of the various lithological types that make up the ore. This difference is factored into the estimation procedure, can drive extraction plans, and results in a much more reliable estimate of the resource and, in particular, the behaviour during treatment and the likely recovery.

UNDERGROUND NUCLEAR EXPLOSIONS

Elliston's ground-breaking research into the formation of igneous-looking rocks had brought in geologists from all over the world to visit Tennant Creek and find out what was emerging. Included in these visits was Professor J G Jaeger, who had been a lecturer of Elliston's at the University of Tasmania but who was now Head of the Research School of Earth Sciences at the Australian National University in Canberra. Jaeger had been commissioned by the UK Atomic Energy Authority to set up a seismic recording station in a remote place in Australia to detect and monitor seismic events globally. As the only operating science group based in remote Australia Geopeko was asked to build and run the station, which it did in an area outside the town of Tennant Creek for many years.

The 20 kilometre-long crossed array of geophones was so sensitive that it could detect kangaroos passing by. It was also able to distinguish underground nuclear explosions from earthquakes, and the staff of Tennant Creek Post Office were constantly entertained by telegrams from Britain to Geopeko's Tennant Creek office reporting the location of "the latest Russian event", or similar.

BETTER DRILLS

Kitching-Warman Hydrodrill 1979

Until 1973 Geopeko had employed contract drillers on its projects, but the need, in Tennant Creek in particular, for well-trained drillers who would cooperate in the tricky business of hitting precise targets at depth with carefully controlled drilling in difficult environments led to the creation of an in-house drilling section there. Geologist Peter Kitto transferred from Jabiru, where exploration had stopped due to the lack of government support, to take charge, in 1973. But as the group's operations expanded it was decided that the division should expand to handle all of the company's drilling: enter Ron Kitching.

At the age of 20 he had abandoned his career as an accountant and travelled to Mt Isa where he began his training in the exploration and mining business and became a driller. In 1953 he and Jack Glindemann founded Glindemann and Kitching Enterprises which became a renowned exploration and drilling company, working all over Australia, including at Tennant Creek. They were an innovative and skilled organisation, and were the first to use wireline drilling and reverse circulation methods in Australia, an attitude that resonated with Geopeko. The company was ultimately wound up and Kitching took up farming in north Queensland for a while, before moving to Rockhampton.

The fact that he might be available came to Geopeko's notice and George Lean flew to Rockhampton to persuade him the join the company. He did so with reluctance, as he did not believe in in-house drilling divisions, claiming that every attempt at creating one had been a failure. It was a rocky association until he brought to the group his ideas for a new type of drill: top drive, dual-purpose exploration rigs. He persuaded Peko to purchase the manufacturing rights to a hydrostatic top drive diamond drill, which had been built for an Adelaide drilling contractor, the late Keith Davis. Peko in turn handed the role of further development to its subsidiary Warman International, who, under the guidance of their chief engineer, Hans Gugger, and Kitching, proceeded to produce the dual-purpose drill rigs. These rigs were the forerunners of the very successful UDR 650s, 1000s, and 1200s which are now commonly used in the industry around the world.

THE SEMINARS

Talk to any Geopeko staff member of the first twenty years and he or she will agree that one of the major contributors to the success of the company was the annual technical seminars.

While working at the Geological Survey of Western Australia Eric Swarbrick and Rob Ryan had been very impressed by the system of seminars that Director Joe Lord had instituted there. Every earth scientist at the Survey had to contribute. The seminars had several beneficial effects: the work of every individual was subject to rigorous scrutiny by their peers; the geologists became more confident and more scientific in their preparation and presentations; and the annual gathering contributed to the teamwork and social cohesion of the group. Having set up the Darwin base they travelled to Tennant Creek to meet the rest of the team and they suggested to Elliston that, with the recent establishment of a team away from Tennant Creek, a similar scheme might be a good idea. He agreed, and the first seminar was held in Tennant Creek in 1967 with Darwin geologists attending. The meeting was a success and it was agreed that it would be held again the following year.

The acquisition of Mount Morgan shortly thereafter meant that Geopeko was about to double in size, with the addition of geologists from a different company who would be coming from a different background and culture, so the second seminar was held in Tennant Creek with the Mount Morgan team attending. One geologist from EZ, representing the Gondwana JV, also attended. It was a riotous affair. The tensions inherent in the coming together of the two groups exploded into a celebration that lasted most of the night and spilled into the town at various stages. Elliston was not impressed but it broke the ice and laid the foundations for the future.

Technically it was a success, and another was approved for the following year at Mount Morgan. The year after that it was held on King Island. Again the get together of the now steadily growing team demonstrated the value of the gatherings and annual seminars were incorporated into company policy.

The ground rules were that all professional staff would have to attend and present a paper on their work during the previous year. No excuses other than hospitalisation were accepted. Drilling programs running at home bases were monitored by telephone if necessary, but with competent and enthusiastic technical staff on the bases that was never a problem. The early gatherings were held at the company's mining operations, and mine staff were invited to talk to the exploration staff about their operation, thus exposing the earth scientists to the other mining disciplines and, more importantly, to the operations that paid for their existence, as well as exposing the mining personnel, who tended to be a bit scornful about the earth scientists in the early days, to the exploration scene. Visits to the mines were an important part of the proceedings, so that the earth scientists could get a feel for what an orebody looked like.

Contributions were also sought from academe on subjects relevant to what the company was doing, and, on occasion, government geologists and joint venture representatives were invited to some of the sessions. In later years, as bases were established in new regions, the seminars followed.

The standard of technical debate was robust. It was encouraged both by the environment of inquiry that Elliston had fostered and by the fact that, thanks to the social side of the gatherings, all present knew each other well and were not inhibited by the occasion.

The seminars served several purposes:

- they exposed the staff to critical evaluation of their work and to ideas and criticism from those with similar experiences and problems;
- they made staff aware of each other's projects so that they could contact each other when common problems were encountered;

- they facilitated transfers of staff between bases by making them familiar with other operations;
- they provided a welcome break from routine and a technical refresher;
- they introduced staff to new ideas and information and developments both from other bases and from universities and government surveys;
- they introduced the exploration personnel to operating mines so that they had a much better understanding of what an economic mineral deposit looked like; and
- they built a team ethos and fostered pride in the company's performance.

Unfortunately, they also cost money. With the retirement of Proud management emphasis changed from ore finding to 'the bottom line' and the seminars were ended as part of a general cutback in exploration expenditure. The message was clear, and a number of senior staff departed. The surviving members tried hard to keep the concept alive, and there were sputtered gatherings from time to time. Williams, by now in charge of exploration, developed the concept of 'budget meetings' which were, in effect, twice-yearly mini-seminars where each member of staff had to present his or her project and justify it. The setting up of the 'specialist groups' was another mechanism introduced to get around the constraints.

These two mechanisms allowed a degree of cross-fertilisation between bases but as Williams has noted "It is remarkable how fences go up between isolated groups and trust and desire for a common goal fall off when people have never met the person they are dealing with in another location". The seminars were the glue that held the company together, both technically and socially. With the glue gone the group began to unravel.

When Peko was taken over by North the problem of merging two very different exploration teams once more arose and a fully-fledged seminar was held in Kalgoorlie in November 1993 along the lines of the original seminars. There were some 50 earth scientists and, as of old, invited guests from the mines, universities, and government.

It was a resounding success and exploration management was criticised for not having held one earlier!

THE MANUAL OF OPERATING PROCEDURES

As the company grew in size, and new bases were opened around the country, it became apparent that some uniformity of practise and procedures was necessary, both so that management could be comfortable with the results of the work that was being done, and personnel moving from base to base, or just joining, had guidelines as to the company's operating practises. The need led to the creation of the 'Manual of Operating Procedures" which became the standard for all bases.

However, recognising that too much of a focus on conformity could stifle innovation, the introduction to the Manual emphasised that it was a guide, not a bible, and the freedom to introduce new practises and techniques was there, provided that they could be justified. As the introduction stated:

> *"The purpose of the manual is to serve as a guide to technical staff in carrying out uniformly the technical affairs of the company.*
>
> *It is not intended to be a set of rules imposed by the Board to govern the activities and methods of technical staff, and specifically the rank of Supervising Geologist in Geopeko Limited carries more authority than this manual. The Board feels that the affairs of the company are better managed by responsible people than by a set of rules and that effective management and leadership does not stem from a book. However, although the individual methods used and decisions taken in the course of management of the*

affairs of Geopeko may be quite adequate and appropriate at the time, the place, and the circumstances, they may not be best for Geopeko as a whole and may introduce a diversity or establish a precedent which is undesirable in the operation of our company.

With decentralized management and the prospect of further decentralized operations in future, it is obviously necessary to have some guidelines. This manual is then intended to reflect the authority of the Geologist in charge in implementing his management practices and technical methods. In its attempt to meet the requirements of the Board for adequate technical guidance and conformity with overall company practices and policy, it will also reflect and be reinforced by Board policy."

The manual also became the basis, many years later, for the creation of the very successful Australasian Institute of Mining and Metallurgy Monograph 12, *'Field Guide for Geoscientists and Technicians'.*

'GEOSCRIPT'

From the seminars there also arose a discussion about a company technical journal and in 1973 'Geoscript' was born. The rationale behind it was to circulate papers of interest to the teams and to keep everyone up to date with staff movements and any other information that might be relevant. The first editorial stated that Geoscript would "comprise a news sheet of Geopeko news, events, notices, contributions from Geopeko staff, discussion of papers, notes of interest regarding competitors activity or developments elsewhere in the industry." This was all pre-internet, in fact pre-fax and in the days when the telex reigned as the wonder machine.

The first Editor was geologist Wendy Hickling, who remembers Elliston saying "that any geologist 'worth his salt' should publish his work and ideas every year for it to be reviewed by his peers." The first editorial, somewhat grandly, also alleged "The view is still held that any geologist engaged in commercial work is not very alert to his science if he does not make at least one scientific observation each year." Consequently all Geopeko professional staff were rostered to provide an article on his or her work, or something relevant and technically newsworthy, each year.

As well as disseminating scientific information, the journal provided a platform for staff to "publish" without going through the minefield of editing and super-high standard of writing and referencing required by the commercial journals such as *Economic Geology*. It was hoped that the authors of the 'better' papers would then present them for publication in formal scientific journals, and many were.

The first issue was keenly awaited, reaching the bases in March 1973. The journal was generally thirty to sixty pages long. Each issue was proudly circulated, with the staff news section consisting of gossip and pranks as well as information relevant to the members of the company. The sources for the gossip and pranks were never revealed. It was said that everyone turned to that section first. The journal provided more of the 'glue' in keeping personnel up to date with happenings in each of the bases, both technical and social.

Geoscript was published four-weekly, to coincide with the thirteen financial periods in each year, from March 1973 till July 1977, then erratically, with a hiatus from July 1978 to June 1979. There were three issues per year for Volumes 6 and 7, five issues each for Volumes 9 and 10. Like the seminars, Geoscript lost all support during the eighties. There was an attempt to revive it starting in 1992 with industry news, reports on mine visits and technical discussions, but sadly few geological papers and little gossip. The last issue known is "The North Exploration Journal" dated April 1994.

SUMMATION

An analysis of Peko-Wallsend's history reveals four distinct phases. From the float of Peko Mines NL until 1975 the company grew steadily on the back of its early mineral discoveries. Failure, in that year, of the new copper smelter at Tennant Creek was the company's first substantial setback.

From 1975 growth stalled, profits fell, and the company entered a period of self-doubt as to its ability as a miner, and there was a questioning of its overall direction. Its next investment was into manufacturing.

When, in 1980, the Ranger One uranium deposit finally cleared the approvals process, the decision was for a public float to fund development. The company's ensuing equity in Energy Resources of Australia ('ERA') renewed its growth in mining but did little to enhance its prowess internally as a miner. Late in this questioning phase, as if to reassert its miner standing, it decided to revisit the Tennant Creek smelter. It would make it work in order to unlock the value of its copper resources which, by this time, far exceeded the remaining gold resources.

The third phase began in 1980 with technical failure of the smelter for a second time. This setback coincided with falling commodity prices, copper (to historical lows which persisted for much of the rest of the decade), tungsten, and gold on a more gradual decline, and coal which started its downward spiral the following year. With all of its mine production affected, the financial consequences were severe. Growth faltered, asset value fell, and profits turned to losses for several years. This can be regarded as the survival phase.

The turning point came in about 1984 with an investment in the Robe River iron operation in Western Australia, giving the company access to a long-term resource as a hedge against its rapidly depleting internal resources.[26] Expansion of its holding in Robe to a controlling interest a year or so later, together with some improvement in commodity prices and a growing contribution from ERA, restored growth and profitability. It was the success of this second growth phase that made the company a take-over target.

The foundation for most of Geopeko's discoveries had been laid early, in the period 1957 to about 1978. In contrast, the sixteen-year period that followed did not see much exploration success, either domestically or overseas.

Some of this fall-off can be attributed to the changing phases within its parent. The instability of the "questioning phase" flowed through as a fall in both board support and funding for exploration. Geopeko reacted by quarantining the scarce available funds for advanced exploration to the detriment of project generation. In its efforts to bring in external funding to bolster exploration it embarked on the technically innovative geophysical survey for the Iranian government (Appendix 2). This brought with it added demands on management and less than desirable attention paid to supporting exploration.

The parent company's 'survival phase' was one of survival for exploration as well. Funds available for exploration were so small that bases needed to be closed, projects farmed out or abandoned, and staffing levels halved. The result was exploration with a much narrower geographic spread, a narrower commodity spread, and operated by a workforce weakened by loss of a good deal of its previous great depth of skill and experience. It was also weakened

---

[26] A philosophy similar to that which drove the original merger of Peko and the Wallsend companies.

by the uncertainty and resultant low morale. But the core Geopeko culture remained. Standards were maintained, teamwork extended to sharing skills between bases, emphasis was kept on field work and drilling, and stability of project funding was re-established. The outcome was a resumption of discovery in the late eighties.

Discovery of a deposit in a new province, or one of a different type in an old province, is normally followed by further discoveries, some quickly by application of the same technique, others later as understanding of the nature of the mineralization advances. Often it is these following discoveries which provide the best financial return to the finder, being bigger or of better grade or simply having better economics by being able to utilise existing infrastructure. Peko Wallsend's growth through to the seventies was largely the result of this type of evolution of discovery in the Tennant Creek field. Such an extension of discovery could well have been expected in both the East Alligator and Rover fields but socio-political issues delayed, and ultimately prevented, systematic exploration and development of the many deposits that were identified.

That probably prevented Geopeko from improving its success record through the seventies and into the eighties. At Parkes, discovery of the Endeavour 22 porphyry copper-gold deposit was followed quickly by discovery of two more, one much larger, which would become the core of the Northparkes Mine. Here the natural extension to search for more was halted in-house by the miners wishing to make the existing resource viable. Exploration was resumed in 1991 after a 6-year gap and resulted in discovery of the Endeavour 48 deposit, the second largest in the field, the following year.

Other factors undoubtedly influenced the change.

- The strong support for exploration from the Chairman and

Board in the early years changed to doubting and questioning in the late seventies and then to something between benign neglect and mistrust through the eighties. That changed back to support under North Ltd in the nineties once integration of the Peko and North exploration teams was complete.

- Contact between the Head of Exploration and the Board, allowing him to act as salesman for exploration and to extoll Geopeko's competence, became more remote over the years and, additionally, the salesmanship did not reach the standards set by Elliston.

- There were restrictions or preferences placed by board and management on the commodities to be sought, based on long-term price trends. That brought to an end promising exploration for tin and for tungsten, and probably rightly so, but it removed some of the flexibility given to the exploration bases in developing new projects.

- There was a condition placed on Geopeko by the Board towards the beginning of the survival period that exploration in the Tennant Creek field was not to be abandoned. Tennant Creek was regarded as Peko Wallsend's home and abandonment would send the wrong message to shareholders during this trying time. Geopeko would have abandoned what it considered was a well explored field in favour of other provinces that it felt offered more likely chances of success.[27] Although the exploration which followed was technically leading edge, and in part led to ongoing research of wide benefit to the industry, the money spent would have had a much better chance of underpinning success if applied to provinces that were less well explored and offered a better probability of success.

- There were preferences developed within Geopeko for deposit types to be targeted, based on anticipated size, grade, and metallurgical complexity. Again this placed constraints

---

[27] In keeping with that philosophy, in 1978 while in charge of operations in the NT Ryan had actually proposed that Tennant Creek exploration be shut down, but his recommendation was not accepted.

on project generation.

- Externally, competition for land use, which developed rapidly in the seventies, continued to grow and diversify into the eighties. The interplay of activism, politics, indigenous land rights, and sensible environmental management saw large areas alienated from exploration and a tightening of regulations governing land use. Confiscation of several Geopeko project areas into national parks or reserves eliminated the likelihood of further discoveries. Lobbying, countering the activists, and compliance reporting took away management time which would have been much better spent supervising and encouraging exploration staff.

What constitutes long-term exploration success?

- Can a small company that accidentally comes across a single huge or rich orebody, thereby generating a very low ratio of value to exploration dollar spent, be regarded as successful?
- If a company, having found or acquired a mine, multiplies its resource base many times by brownfields exploration, can it be egarded as a good explorer?
- Can a company be said to be successful if it consistently identifies, by shrewd analysis of the market, the potential of finds by smaller companies as of major importance and buys in, thereby showing a successful ratio of dollars outlaid to resource value?
- Is a 'technical success", where, by good exploration, a company finds several mineral bodies that are not economic at the time but subsequently become so, an arbiter of that company's success?

It is proposed here that a company can be judged to have been a successful explorer if it:

- finds orebodies in a range of greenfields provinces over a significant period of time;
- finds them in provinces of different ages and geological

characteristics;
- finds different ore types and different commodities; and
- finds them at a comparatively low exploration cost.

An organisation may be said, also, to be technically successful if it operates at a high level of technical expertise and makes important contributions to the understanding of ore geology and the practise of exploration.

Luck, particularly being in the right place at the right time, plays a role in most discoveries although it can be argued that being in the right place at the right time again and again is a measure of a company's competence, even though it may look like luck.

How does Geopeko stand up against this list?

1. Mineral deposits, not all of them economic at the time but many developed later as market conditions changed, were found in the Warramunga Province, the Pine Creek Geosyncline, the Lachlan Geosyncline, the New England Orogen, Archaean greenstone belts, the Bangemall Basin, the Hodgkinson Basin, the Georgetown Block, and the Quaternary. There were discoveries, in the forty years from 1954 to 1994, in NSW, the NT, Queensland, Tasmania, and Western Australia (Appendix 5).
2. Discoveries included precious metals, a range of base metals, uranium, and mineral sands, in porphyry systems, in various types of vein deposit, in IOCG systems, in skarns, in VMS environments, and in dune systems.
3. The first orebody was found in 1957 and the last (before the company ceased to exist as a stand-alone operator) in the early nineteen nineties.
4. As the McKinsey Report on exploration success has demonstrated, Geopeko found these deposits at costs comparable with the industry and, in terms of value of discovery per exploration dollar over the period of that report, better than any of its competitors.
5. Through this period it was at the forefront of developments

in exploration science and technology.
6. The company's operating philosophy, of integrating geoscientists with various disciplines into exploration teams, has established for the industry a proven exploration management model.
7. At Geopeko's instigation Peko-Wallsend was involved in the establishment of university positions, in funding research, in funding research scholarships, and in providing student and graduate employment.
8. In the work of John Elliston, strongly supported by Peko-Wallsend, the company has made a major contribution to the understanding of rock and ore formation. Many geoscientists who started their careers in Geopeko have gone on to be world-class researchers (e.g Sue Golding, Ross Large, Terry Lee) and world-class explorers with other companies. Many, having left Geopeko, have enjoyed success in ore-finding projects of their own.
9. Developments in instrumentation and exploration techniques by the company and its associates were in some cases innovative and ground-breaking.

By all of the criteria set out above, therefore, Geopeko can be regarded as having been a successful company.

The factors that influenced it's performance, and the reasons for the its success, are generally thought to be as set out below:

- The steady support of the Chairman of the Board, John Proud, who carried the Board with him, and the willingness, in the early days of the Peko group when times were tough, to give priority to exploration funding at the expense of dividends.
- Early and continued exploration success which justified that initial support and strengthened the Chairman's case;
- Lew Richardson's expertise and support for exploration, his insistence that the 'specialists' should operate as an integral part of the exploration team, not as 'consultants' to be

brought in when thought necessary by the geologists, and his influence with Proud and the Board;

- as the head of exploration, John Elliston's strong advocacy for exploration at board level. Geology and exploration technology are very specialised disciplines that non-scientific people have difficulty grasping. This is not a criticism, merely a fact of life. Without competent and informed, and persistent, advocacy at the top levels of decision-making, exploration will usually lose out to those matters that the directorate is more comfortable with, such as "the bottom line", the purchase of proven assets, or OH&S. It is no accident that most successful ore-finding companies, according to the criteria laid out here, are either run by earth scientists and mining engineers, or at least have strong and influential representation of both at Board and senior management level;

- the high professional standards set and the scientific rigour demanded by Elliston and Richardson and supported by their line managers. This applied not only to the scientific aspects of work but to logistics, tenement management, dealings with government and the public, and reporting, and included the para-professional and technical staff at all levels;

- mentoring of junior staff by experienced senior staff. This involved many technical and scientific aspects.

- the policy that detailed work should only be carried out on ground held under secure tenure;

- the sense of continuity so that project staff could progressively collect and interpret information towards a thorough understanding of the geology and the attributes around which exploration tools could be applied;

- a rigorous analysis of each drill hole result to test whether it had done what it was supposed to do, or to explain why it had not, and the willingness to drill enough holes to test a prospect until the answers had been found;

- cross-fertilisation with mine staff and exposure to mining operations, which allowed the exploration staff to understand the requirements for converting interesting discoveries to economic reserves and mine managers to gain a better appreciation of the exploration effort;

- good leadership, which encouraged technical performance and engendered team spirit, and a very flat management structure, with base and division leaders reporting directly to the head of exploration;

- regular visits to operations by senior management with the intent of positive interaction, encouragement, team building, and streamlining operations.

- avoidance of any semblance of "command and control" philosophy;

- the sense of security of tenure and company support for the field crews that pervaded the company;

- the constant search for new geological concepts and exploration methods and techniques, and the willingness to adopt them;

- the excellent technical support provided by the specialist divisions as the company's operations expanded, boosted by the stationing of specialists on bases where they were involved daily in the base's activities;

- exposure to other's ideas by way of the seminars, the in-house journal 'Geoscript', visits to group operations, and input from tertiary institutions and think tanks. Exploration personnel obtained a rounded view of, and an appreciation of, the industry in which they worked;

- focus on regions with good geological potential rather than trying to outguess the market with regard to commodities likely to be in demand;

- a focus that was on understanding ore-forming processes which led, in turn, to the selection of regions with high potential for discovery as exploration targets;

- the establishment, in the early phase, of exploration bases in, or as close as possible to, target geological provinces rather than in capital cities or larger towns, which meant that there were small, coherent exploration teams operating as tight units with good communication and professional support that were in close touch with what was happening in their target provinces;

- the freedom given to local managers both scientifically and logistically;

- the inclusion and training of the para-technical staff in the company's work, as it raised the quality of the routine day-to-day tedious data collection which, if done badly, can severely compromise a field program.

In summation, it can be said that Geopeko's early history is to a great extent a text-book of how mineral exploration should be conducted. Its success rate and costs testify to that. The overall history of the group also highlights the value of the correct management of the company. As the group grew, and as more and more of the decision making about what to look for, and where, became centralised in head office and at board level, remote from the field crews, so the success rate fell away.

This was partly due to cuts in funds. Justified or not, they happened and impacted on morale, but there is also the perception, when reading the story of the latter years, that the input from the exploration teams was more and more ignored and that morale was consequently sagging.

If an exploration team is to succeed it must have the confidence of the directorate, and must know that it has, and must be sure that when it has a project to recommend it will be listened to.

But it must also justify its existence with success.

APPENDIX 1:

EXTRACT FROM A REPORT BY MCKINSEY & COMPANY, INC[28]

SUCCESSFUL MANAGEMENT OF MINERALS EXLORATION IN AUSTRALIA

REPORT TO SURVEY PARTICIPANTS

Nearly $100 million goes into minerals exploration in Australia each year. Many individual companies invest over $1 million a year, and a handful of companies as much as $4 million or $5 million. For some companies, the returns have been extraordinary; for others, the investment a failure. Little has been written about why some companies succeed where others fail. Can the right management approach affect the outcome, or is it simply a manner of luck in a high stakes game of chance? This report is an attempt to find some answers to the question.

In March this year, McKinsey & Company undertook a survey of management approaches in minerals exploration as part of a project for Conzinc Riotinto of Australia Limited. Fifteen companies were asked to participate in the survey and 13 agreed to do so. The companies were selected to give a mix of highly successful, moderately successful, and less successful efforts; a wide range of spending; and a variety of management practices. We agreed to provide participants with the findings of our survey in return for their cooperation with us in discussing in detail their approach to exploration.

---

[28] By permission of Rio Tinto Limited.

Interviews in an area as subjective as this do not provide much hard data. However, they do provide strong qualitative impressions. Where we thought we saw a difference between the approaches used by successful explorers and those used by less successful explorers, we have singled it out for discussion in this report. We were aided in our interpretation by fairly extensive work carried out in other parts of the McKinsey organization on the management of oil exploration and on the management of technology generally.

We should note that they're the variety of different strategies and approaches which have proved successful. We do not want to imply that there is only one way to find ore bodies. Rather, we are attempting to understand what strategies and management approaches are most likely to be successful.

Most people recognize its success in managing expiration is different from success in management of other things. But what makes it different? Let us start with our definition of "success".

For the purpose of this survey, we have define success in exploration in three ways:

J Highly gross little of value heard a lot of expiration

f Low exploration cost per find

J High ranking by peers in response to the question, "who are your best competitors?".

Not surprisingly, the above three measures have a fair degree of correlation. This makes it possible to distinguish between the more successful and the less successful companies with a fair degree of confidence. Since we were mainly interested in success, our sample of companies tends to be biased towards ones who have been reasonably successful (Exhibit 1).

What about luck? Is it possible that the pattern of finds over any period can be explained simply as a random phenomenon, related only to how much money the company invests in exploration... the more money you throw at exploration, the greater your chances of success? Clearly, there is some relationship between dollars spent and odds of success. However, our analysis of past data shows that several companies are improbably successful - far more successful than a statistical distribution based on dollars spent would predict. By contrast, a fair number of companies are improbably unsuccessful. We conclude that the dice are loaded against any management who treats the exploration question mainly as luck.

EXHIBIT 1

Measures of success

Participants	GROSS METAL VALUE/EXPL. $	COST/FIND ($ MILLION)	RANKING BY PEERS
A			
B			
C			
D			
E			
F			
G			
H			
I			
J			
K		∞	
L		∞	
M		∞	

APPENDIX 2:
THE AUSTIREX STORY

The discovery of Ranger One, and the ensuing exploration in the East Alligator River region, was the first serious involvement by Geopeko in uranium exploration. The discovery was a spectacular example of the benefits of airborne regional geophysical surveys, specifically airborne radiometry. The survey was conducted by contractor GRD Surveys and managed by L A Richardson & Associates.

Following the discovery Geopeko became engaged in a very active program to define and understand the Ranger 1 orebody and uranium mineralisation, and to explore the large surrounding Munmalary tenement area for repetitions of Ranger 1 style mineralisation. Richardsons was thrown onto a very rapid learning curve on nuclear physics and its application to radiometric measurements and uranium exploration.

The work indicated that there was more to an airborne radiometric survey than simply as an "anomaly finder" – that it was important to have good positioning, precise discrimination between Th, U, and K, and to have data that could be reliably and accurately plotted and presented in two dimensions, not just along-line.

As the team progressed up this learning curve it began to really appreciate the importance of data quality. That meant developing a much deeper understanding of the issues of positioning, corrections for atmospheric radon and cosmic radiation, spectrometer stability, pre and post flight tests, calibration procedures, altitude and parallax corrections, and better ways to present the data. Rather than just accept what contractors would provide, Richardsons insisted on getting "inside" their systems and procedures, ensuring that they knew exactly what and how they were measuring and what they did with the data.

Naturally, much of what was learned about airborne surveys was put to good use on other projects, such as at Tennant Creek NT, Mt Morgan Qld, and the Tullamore Syncline NSW. During this period contractors used included GeoMetrics, GeoSearch, or Aero Services for flying, and Pittmen Data Systems for processing. A good working relationship was developed with Geosearch, led by Kevin Radford, and Pittmen, led by Bob McKenzie and Ian Campbell and both were involved in the contract with Iran.

Geopeko was then encouraged by the Australian Government, Overseas Trade Department, to seek work overseas. Overseas Trade proposed that Geopeko team up with WMC to seek work in Iran.

Iran's efforts to develop a peaceful nuclear energy program date back to the 1950s, when Shah Mohammad Reza Pahlavi was in power. Iran, then an ally of the West, heavily relied on assistance from the United States to gain access to nuclear technology. In 1957, Tehran and Washington reached an agreement on cooperation in research on the peaceful uses of atomic energy under the American "Atoms for Peace" program. In 1973, a study carried out by the Stanford Research Institute in the United States recommended the construction of atomic power plants capable of generating 20,000 megawatts of electricity in Iran before 1994. Soon afterwards, Iran set the goal of producing some 23,000 megawatts of electrical power from a series of nuclear power plants over the next two decades. The aim was to generate needed electricity from nuclear power as opposed to oil, thereby freeing up crude output for export and refinement. To this end, the Atomic Energy Organization of Iran (AEOI) was established in 1974, and mandated with implementing the aims of the Iranian nuclear energy program.

AEOI signed a contract with the Iranian firm Uriran to oversee the exploration of potential uranium deposits. The exploration was to cover a span of some 600,000 square kilometres, or over a third of Iran's surface area, and requiring approximately 960,000 line kms of flying.

Visit to Iran – July 1975

In anticipation of likely meetings in Iran, Geopeko rapidly put together brochures and documentation to outline its capabilities and support the marketing efforts. On July 23rd 1975, an Australian delegation consisting of Stephen Lacher (lawyer with N.G.Green and Co), Rob Ryan, and Bob Richardson (Peko-Wallsend), and Eric Cameron (WMC) visited Tehran for meetings with Reza Niasmand, Chairman of Uriran. This visit was at the request of the Australian Government to discuss the possibility of the two companies providing technical assistance to Uriran for the exploration for uranium. A document entitled "Preliminary Statement of Corporate Histories and Summary of Exploration Capabilities of Joint Australian Delegation of Geopeko Limited and Western Mining Corporation" was prepared for the Atomic Energy Organisation of Iran, along with brochures about the respective companies including "Geopeko Limited – An Australian Exploration Enterprise".

Niasmand asked the delegation to prepare an hypothetical exploration program for uranium to demonstrate our knowledge and experience in uranium exploration. So Cameron, Ryan, and

Richardson put their heads together and produced a document entitled "Model Programmes for Uranium Exploration in Iran" which was completed in August and presented to Uriran at a later visit to Tehran in February 1975. The document outlined exploration programs for three specific deposit types – vein type, sandstone type, and calcrete type.

Uriran then advised that it was not interested in accessing a wide range of exploration services, although it might be later. The initial requirement was airborne geophysical coverage over selected large areas of Iran, and they would be seeking bids from survey contractors.

Dilemma! The team had a thorough understanding of the technology and the requirements but no aircraft, no airborne systems, no pilots, no data processing software, and no plotters.

Things moved somewhat slowly while Uriran built its own organisation. An Iranian geophysicist Sohrab Batmanglidj was appointed as CEO to manage the bid process and the ongoing survey contracts. Sohrab was educated at the Colorado School of Mines and resident there at the time. Obviously the Geopeko team was keen to meet him and present their credentials, so a delegation of WMC and Geopeko people met with him and his advisors in Denver, USA, in December 1975. It became apparent as discussions proceeded, both in Denver and Tehran, that the proposed surveys were going to be very extensive and that the Iranians wanted high quality data - "the best that western technology could provide".

An Australian geophysicist, Terry Crabb was appointed by Uriran to supervise the contract. The team were somewhat apprehensive because they had never heard of him and were keen to meet him. As it happened he was arriving in Sydney on the day of the Richardsons Xmas picnic at Chinaman's Beach, so he was invited along. He turned out to be very easy to get along with and fitted in very well.

At about this time, WMC got cold feet about the venture and withdrew. They did not have the same level of expertise and knowledge in airborne geophysics and were losing track during discussions of the technical issues. Peko and LAR then incorporated the company, Austirex Aerial Surveys Pty Ltd, to be the lead contractor. It was owned by Peko- Wallsend 87.5% and LAR 12.5%. The board consisted of Bob Richardson, Ron Moore, and John Elliston. John was Chairman and Richardson was Managing Director. The name "Austirex" was constructed from "**Aust**ralian-**Ir**anian-**Ex**ploration". Austirex then set about engaging Geosearch and Pittmen Data as sub-contractors for the flying and data processing. Then began a frantic and intense period of negotiation with sub-contractors, further discussions with Uriran about technical issues and survey specifications, examination of some thorny technical problems that had to be solved, aircraft type and system design, legal and commercial issues regarding doing business in Iran, Austirex staffing, and preparation of a bid submission.

Several more meetings were held with Uriran in Tehran (Sept 1975) to discuss technical aspects of the job. It was during this period that the Austirex team became aware that it was in the forefront of the technology compared with other tenderers, and that the relationship with Uriran was developing into a constructive process with Austirex having a significant impact on the design of the survey specifications. Clearly Uriran respected the input and began to understand that Austirex was not approaching this job just as survey contractors, but as experienced exploration geophysicists with an understanding of the exploration objectives of the survey. This developing relationship had important implications for the bid submission, and was a major factor in Austirex winning a contract, with a very high bid price. Uriran wanted Austirex there "to set the standard" for the whole survey program.

The survey program was to cover an area of approximately 600,000 km^2, or about one third of Iran, over areas that had been selected on the basis of their prospectivity for uranium. The country ranged from flat and featureless to rugged and mountainous. Clearly it was going to require both fixed wing and rotary wing aircraft.

The basic specifications called for were as follows:

- line spacing: 0.5, 1.0 and 2.0 km;
- tie line spacing: Ten times line spacing;
- spectrometer: 256 channel, 50,000cc NaI detector;
- magnetometer: 1.0nT sensitivity;
- mean terrain clearance: 120 m;
- sample interval: 1 second; and
- maximum aircraft speed: 70 m/sec.

The data were to be presented in the following forms:

- digital data on magnetic tape;
- analogue strip chart records of all survey flights, test lines and calibrations;
- computer plotted stacked profiles at 1:250,000 scale of the following geophysical parameters, total count(c/s), K(c/s), U(c/s), Th(c/s), Bi-214(air), U/Th ratio, U/K ratio, Th/K ratio, total magnetic field, and radar altitude;
- multi-plot profiles for every flight line, showing all the above parameters;
- computer-plotted contour maps at scales of 1:50,000 and 1:250,000 of the following geophysical parameters; total count, uranium, uranium equivalent(ppm), thorium equivalent(ppm), potassium equivalent(percent), and the

ratios U/Th, U/K, Th/K, and total magnetic field; and
- computer-plotted flight path maps.

All of the plotted maps and profiles were to be micro-fiched. For each map area there had to be 10 individual maps, so that a total of more than 5,000 individually machine-plotted maps was required.

Crucial to the success of the project was the appointment of a General Manager to take charge of Austirex in Tehran. Richardson was determined that the appointment be a geoscientist with a state-of-the-art understanding of radiometry, and someone that was prepared to take a key role in a venture that was on the leading edge of the technology, so he pursued Phil McSharry, who he knew from their involvement in ASEG (The Australian Society of Exploration Geophysics) and with whom he had jointly run a workshop for members on radiometry.

McSharry's M.Sc. thesis in 1974 was entitled *"The collection and processing of airborne radiometric spectrometer data"*, and in Richardson's opinion he knew more about the science than anyone else in Australia. He was Supervising Geophysicist with Getty Oil at the time. It didn't take him long to become captivated by what Austirex was doing and the opportunity to be involved, and he joined LAR in January 1977, to be seconded to the Austirex project as General Manager in Iran.

Particular technical issues that had to be solved included:-

- corrections for the effects of atmospheric radon;
- spectrometer stability;
- requirement for a very high sensitivity spectrometer, with a sensitivity of 1.0ppm eU;
- management and recording of a massive amount of data every second; and
- height correction procedures.

McSharry had the lead role in solving these issues, in collaboration with others especially Ian Campbell, Rod Tuson, and Albert Berkavicius. Phil also spent some time working with Bob Foote in Dallas, Texas working on spectrometer calibration procedures and other processing issues. Ian was the software developer with Pittmen Data, Rod Tuson was a surveyor engaged in geophysical work with Richardsons, and Albert was an electronics engineer and system developer working with Geosearch. All three were geniuses in their field.

The Iran survey program was divided amongst three successful bidders – Austirex, Prakla Seismos (German), and CGG (French), the latter two large and long-standing international contractors. Prakla and CGG each had areas that required both fixed and rotary wing flying, whereas the Austirex areas were all fixed wing because Austirex used the more agile NOMAD aircraft.

In late 1976 Austirex signed a contract for 315,000 line-km of fixed-wing flying and map production. The total contract price was US$15.5 million, or approximately US$50 per line km. The areas allocated to Austirex are shown in blue on the accompanying map.

Insurance was taken out with the Export Finance Insurance Corporation (EFIC). In the event of non-payment by AEOI, or non-reduction of a Bank Guarantee, for a non-delivery of maps, after sixty days Austirex could terminate the contract and be paid out by EFIC.

Prior to this job, Pittmen was a small organization of three very talented people – John Pitt, Bob Mackenzie, and Ian Campbell. They changed their name to Infographics Pty Ltd and put on more staff. Richard Collins was engaged as manager of the Tehran operation and moved there with his family. Three Data General mainframe systems were bought along with all the necessary peripherals - one for Tehran and two for their Gordon office. They packed up the mainframe system that was destined for the office in Tehran and dispatched it through a reputable freight company. It never arrived, at least not to the Davoodieh office in Tehran. Despite extensive enquiries and many trips to freight warehouses in Tehran, where there were endless piles of lost and homeless freight, the team gave up and ordered another system. Many months later they found out that it was sitting in Esfahan University, some 360kms south of Tehran. It then found its way back to its rightful home in Davoodieh, thanks to some careful negotiation of Iranian practices.

The Austirex airborne geophysical system was the result of combining practical exploration experience with advanced engineering design. In terms of sensitivity, ease of operation, reliability and data quality it was the most advanced of its type in the world at that time. The whole system, including aircraft and peripheral devices, cost approximately $1million (1976 dollars). It was installed in Perth by Geosearch engineers and technicians.

The spectrometer used was a GeoMetrics GR-8000 combined with GeoMetrics 50,340cc main pack detector and 8,390cc upwards looking detector. The large gamma ray detector improved the signal-to-noise ratio of the radiometric data and was able to reveal the expression of smaller, weaker anomalies. The improved count rate statistics enabled subtle contrasts in radiometric signatures of rock-types to be enhanced and mapped reliably across line to line. Gathering 256 channels of main detector data allowed better quality control of the data: spectral instability, electronic interference, and the presence of stray radiometric sources could be identified and corrected retrospectively. The reliability of the data could thus be guaranteed and verified. The continuous measurement of radiation from atmospheric Bi_{214} and cosmic sources enabled the data to be corrected for radiation and contribution from non-terrestrial sources. By rigorous processing, and reference back to calibration flights over geochemically sampled areas, the data could be reduced and presented as profiles and/or contours of equivalent uranium (eppm), thorium (eppm), and potassium (percent). That allowed survey data to be quantitatively compared in absolute terms from one area to another.

The main detector consisted of three 16,780cc thallium-activated NaI crystal packs coupled to a 256 multi-channel analyser. The packs were manufactured by GeoMetrics in Palo Alto California. Each pack consisted of four 4x4x16 inch crystals each optically coupled to a photomultiplier tube, and the entire array was housed in a shock-mounted thermally insulated container with its own temperature control.

The contract specifications called for a means of monitoring and correcting for atmospheric Bi_{214} which is a strong gamma emitter. Bi_{214} is a daughter product of radon gas (3.5 days half- life) and the amount of radon gas in the air varies significantly from day to day. Radon emanations from the ground vary according to the local geology, climate, and variations in atmospheric pressure. The Bi_{214} adheres to atmospheric dust particles and hence gives rise to spurious gamma radiation from above the aircraft, contributing significant "noise" to the terrestrial gamma signal. Pre and post check flights, and daily flights at high elevation, could go some way to monitoring and correcting for this effect, so it was decided to go for the ultimate solution – an upwards looking crystal detector. That required some research into the optimal detector size and configuration. The system that was adopted consisted of one 8,390cc NaI crystal detector pack sitting on a 2.5cm thick lead slab shield over one of the downwards looking main detector packs and coupled to a 128 multi-channel analyser. That could monitor radon effect in real time every second.

Using a pulse height discriminator, the multi-channel analyser sorts the pulses from the photomultiplier tubes into 256 channels. The amplitude of each pulse is proportional to the energy of the gamma ray photon that caused it, thus the system measures a gamma ray energy spectrum for each second. A 128 channel spectrum is also measured for the upwards looking crystal, so the system has to cope with a massive quantity of data every second. Issues such as the effect of pulse dead time become important.

The development of this system was a joint effort between Canadian based company Sonotek Limited and GeoSearch's brilliant engineer, Albert Berkavicius. It was a software-controlled system built around a Fabritek MP12 computer.

Two DIGI-DATA 1600 tape decks were installed in the system. Two recorders were used, to minimise tape changing during survey. Analogue records were recorded on a GeoMetrics GAR chart recorder. Channels were Mag 0-100nT, Mag 0- 10,000nT, Total Count, Thorium, Uranium (stripped), Potassium (stripped), and altitude.

The magnetometer system used a proton precession type Varian V85 toroidal omni-directional sensor and front end electronics, with Austirex modifications. The sensor was mounted on a five metre-long fibreglass "stinger" at the rear of the aircraft. Processing of the magnetometer precession signal and conversion to magnetic field strength was done by the Sonotek/GeoSearch acquisition system. Sensitivity was either 0.1nT or 1.0nT depending on the cycle times.

The navigation system employed both Doppler radar and tracking camera. The Doppler radar incorporated a Decca type 72 Doppler system, a Sperry GM9 Gyro platform, and a Decca TANS 9447D navigational computer. The tracking camera was a Vinten MK III 16mm camera operated by the Sonotek master timer.

The choice of aircraft was a critical decision with important ramifications for the bid price, and the design of the system. An aircraft was required that could carry a heavy system and at least three people, fly low and slow, be nimble and powerful enough to cope with the rugged topography, with adequate range for long flights from the operating base, and offering high reliability in a challenging and often remote environment. Several alternatives were looked at, including the Fokker Friendship, Skyvan, and DC3, but finally the Australian-built Government Aircraft Factory NOMAD 22B was selected. It was an excellent choice and gave Austirex a distinct advantage over its competitors. It was powered by two Allison turbine engines developing 400HP. It had a roomy cabin for an uncluttered and accessible system installation.

Cruise range was 1,352 km. It had a slow stall speed of 91 km/h and was sufficiently agile to handle a lot of the more rugged country, that otherwise would have required a rotary wing platform. Two of these aircraft into service, equipped with identical geophysical systems, were ultimately put into service.

The survey was managed from a base in Tehran, which had to house a number of functions – technical management, preliminary data processing, flight path recovery, management of supplies to the field bases, liaison with Uriran, housing of some personnel, compliance with Iranian civil and military regulations, etc. A five story apartment block was rented in Davoodieh, a suburb in NE Tehran. Two floors were dedicated to the Infographics computer system and their personnel, and the rest to flight path recovery, offices, and accommodation. Austirex also appointed an Iranian as the commercial manager at Davoodieh, and a retired Iranian army colonel ("the Colonel") to assist in liaison with military and government officials.

Flying began in September, 1977, in an area near Tabriz in north-western Iran. In addition to survey flying there was an extensive programme of test flights to be carried out to test and calibrate the system, and to derive parameters for subsequent processing of the data. As some of this information was not required until processing commenced, some of these test flights were conducted quite late in the programme. Some flights were conducted over Lake Rezaiyeh in NE Iran, at Isfahan and Shahdasht, and for system calibrations over test pads at Ghale Morghi airport. The test pads construction had been supervised by US radiometrics expert Bob Foote. There were five pads with varying quantities of Th (8- 46%), U (2-18%) and K (~28%).

Particular test flights and checks included the following:

- cosmic radiation v altitude tests over land and water for cosmic background;
- altitude attenuation flights;
- parallax test flights;
- heading error tests;
- flights for roll and pitch corrections;
- test pad calibrations of the spectrometer; and
- tests to determine geometric factor for the UP/DOWN detector array.

Navigation and flight path recovery proved to be a nightmare. The survey areas had generally featureless terrain and the quality of the mapping and position control was poor. In more advanced countries, such as Australia, the mapping quality is generally good enough to get adequate flight path fixes using tracking film to plot onto controlled photo mosaics. These days position can be established reliably down to metre precision using GPS-based systems, but there was no such facility in the nineteen seventies.

The first positioning issue to solve was the provision of maps for flight path control – maps that would enable the pilots to navigate and fly lines within specification. Austirex had been assured that aeronautical maps at 1:50,000 scale would be made available by the Iranian government but when the time came the military vetoed it and Geosearch had nothing to work with. It then had no option but to fly on lines of constant bearing - in effect Rhumb lines. A Rhumb line is a line crossing all meridians of longitude at the same angle, i.e. a path derived from a defined initial bearing. That is, upon taking an initial bearing, one proceeds along the same bearing, without changing the direction as measured relative to true or magnetic north. To calculate and plot these lines on a map, it is necessary to take into account the geodesic curvature of the earth. Ian Campbell devised a method of calculating rhumb lines based on a moving approximation of the earth's curved surface and was able to supply plotted flight path control for Geosearch.

Flight path recovery was then another headache. The system that was adopted consisted of a combination of on-board Doppler radar and tracking film recovery onto available topographic maps, which were ultimately made available. Doppler radar measures cross-track and along-track velocity. By integrating velocity the system outputs cross-track and along-track distance travelled. However the errors accumulate with distance travelled and regular fixes back to absolute position were needed. The only way to do this was to scan the tracking film images (taken every second) and try to identify features on the available maps, and also tie-line and flight-line crossovers. This enabled the position information provided by the Doppler system to be regularly corrected and tied back to known absolute position. Identifiable features were few and far between, so it required flight path recovery personnel with exceptional skills in pattern recognition. As the kilometres of data poured off the NOMAD the production of digital flight path data became critical and the team rapidly got behind. Ultimately this was solved by putting together a large team of skilled and dedicated personnel working flat out on this work, in Tehran and later in Sydney.

Navigation and positioning raised a serious security issue with the Iranian military. They were concerned that the survey could be taking photos of strategic military installations, missile silos, airports, etc. with the tracking camera. They initially insisted that a military observer be carried on all flights, but after a few long flights at low level the appointed observers turned a shade of green and decided that it was OK just to stay in the camp, drink Coca Cola, and check the tracking film when the plane returned. But that also had a few teething problems: when the film from the first few flights was returned to the Austirex crew in a box, cut into many small pieces after they had scissored-out sections that they thought may have been of a sensitive nature. That was soon sorted out – no continuous tracking film, no survey.

Despite all the assurances by Overseas Trade people in 1975 that Iran was stable and that the Shah had iron control of the country, there had been simmering opposition to him. Many of the his reforms, especially those involving women, infuriated conservative Muslims, led by Ayatollah Khomeini, a Shiite scholar. There was also opposition from a middle class that sought greater political freedom. The opposition boiled over and brought millions of people onto the streets in late 1978.

The situation became dangerous for the Austirex staff so work was shut down and everybody got out of the country on 16th December 1978. The Shah and his wife fled in January 1979, ushering in a brief period of confusion before Khomeini assumed control as Supreme Leader over what became the first Islamic theocratic regime in the modern Middle East.

The Davoodieh office and the computer hardware were left in the care of the Iranian staff. The aircraft were parked and covered at the airport at Zahedan in the south-east of the country, and data processing continued in Sydney. At that stage the flying was approximately 80% complete and 270,000 line km of data were ready to process and turn into maps.

Kevin Radford and Bob Richardson went back to Tehran in mid 1979, crated up the computer systems and organized for them to be shipped back to Sydney. Kevin and another pilot went on to Zahedan and flew the Cessna and the NOMAD out of the country to Karachi.

Concerns about delays, and the risk of relying on only one aircraft and system to complete the survey on time, had led to the purchase by Austirex of a second NOMAD and fit-out in 1979. This aircraft was used in Australia and did not do duty in Iran.

Most of Infographics' previous projects were small scale compared with the Iran project. Given the massive amount of data and the processing and plotting required, it became clear to Ian Campbell that there would be no opportunity for manual intervention in the processing and plotting production line. The whole process from the delivery of flight data tapes to map production had to be automated. Ian (at that stage Ian alone) had the knowledge and the foresight to realise that if they tried to manage this job using their existing software packages it would quickly devolve into a shambles. So he set about building a new processing system from scratch, with a system architecture that would enable all the elements of the processing to proceed automatically from raw along-line data through to final profiles and grids to plotted output in accordance with the Uriran specifications.

Austirex was paid only on map delivery, so until Infographics completed their processing system and started producing maps, it was not being paid and was running into serious financial trouble. The whole success of the contract was resting on one man's shoulders and Ian was under huge pressure. It is a testament to his intellect and emotional toughness that he was able to concentrate on the job and produce a processing system that was probably the best in the world at that time, and finish the job.

One issue that was particularly frustrating was the necessity for automatic "spike" removal from along-line magnetic data. For normal size jobs this could be handled by operator intervention, but this was not an option on this job. Ian developed some very clever automated routines using algorithms based on a method of "differences" that tested the validity of each spike and either accepted or replaced it.

Meanwhile the rest of the team was doing its best to convince Peko management that this was the only course to follow, and giving Ian as much support as possible. PWL CEO Don Stewart was taking a keen interest, no doubt largely because of Austirex's deteriorating financial situation and the possible effect on Peko's bottom line. At one meeting he lined Richardson and McSharry up against the bar and left with the following advice – "Now listen you two, there are three little words I am not afraid to say – they are 'you' and 'are' and 'fired'." But he did understand the situation and at a subsequent meeting Ian convinced him that this was the only way forward and that he had the ability to finish the job. Ian had other pressures at the time, particularly an ongoing dispute with builders who were constructing his house. Stewart pulled in Peko engineers and lawyers to fix the house problems at no cost to Ian.

When maps finally started pouring off the plotters in September 1979 Stewart dropped in to witness the event. It was congratulations all round and Stewart asked Ian to "pop down to my car and you will find a bottle of Bollinger in the boot". Ian arrived back with the whole case of Bollinger and things got fairly messy, culminating with breakfast at Balmoral beach at 7am.

With the Iranian work in production it was evident that there was a wider future for Austirex as an international airborne geophysical and data processing contractor beyond Iran. Large survey contracts in Saudi Arabia, SE Asia, and the USA were already under review, as well as the usual run of smaller surveys in Australia. Geopeko was also keen to have ongoing access to the Austirex capability for its exploration activities. So began a lengthy period of discussions and protracted negotiations between June 1979 and early 1980 to see how the capabilities and resources of Austirex, Infographics, and Geosearch could be consolidated into one organization. Without going into detail, there were some thorny commercial issues to resolve. Geosearch and Austirex each owned a Nomad aircraft and system. Several ways of combining the organisations and assets were considered. Suffice to say that Austirex finished up with both Nomads and systems, and the Infographics people and systems, and "Austirex International" came into being as a real force in the airborne geophysical contracting business.

After the revolution, the contract was in Force Majeure. Sohrab Batmanglidj had left Iran for the US. It was learned later that he had been shot in the arm by a revolutionary guard and was receiving treatment in the US for severed nerves. Austirex was now dealing with a Mr Nowroozi of AEOI. Despite the unrest, AEOI made it clear that they wanted to receive all of the maps. By October 1979 approximately half of the final maps and profiles had been delivered. Map production then slowed down and rolled on through 1980. The team was reluctant to risk sending these very valuable maps by normal airfreight so they were hand carried by Austirex people to Tehran. In late 1980 Austirex completed all the data processing, and map production, for data that it had managed to gather before the revolution, and was paid out for the residual by EFIC. The final report to AEOI was completed and delivered in 1985.

Austirex was the only contractor to finish the job, or at least what it was possible to finish. The last 15% of the flying could not be completed because of the revolution, but would have been completed otherwise. Tragically, in December 1977 CGG crashed their DC3 in Iran with the death of all on board. Prakla had trouble meeting the specifications, had operational difficulties, and were unable to finish.

Assessment of the value to Geopeko

On the one hand, Austirex was a serious distraction from the main business of Peko- Wallsend and Geopeko. It caused considerable stress and strain on the organization, at all levels.

On the other hand;

- it gave employment for a large number of people within the Peko-Wallsend group and its subcontractors;

- it demonstrated to the wider industry and Geopeko staff that the company was on the cutting edge of airborne geophysical technology and could compete with the world's best in this field;

- the Iran contract contributed immensely to the knowledge of airborne geophysics and this knowledge flowed on to Geopeko's other projects. Many projects received essential guidance from airborne geophysics including Parkes, NSW; Southern Cross, WA; Peak Hill, WA; Elliot Bay, Tasmania; Herberton, Qld; Pine Creek Geosyncline, NT; Coen, Qld; King Island, Tasmania; Rover, NT; and Woolner, NT. In 1984 Geopeko contracted Austirex to re-fly the whole Tennant Creek Central Field for high quality magnetometry and radiometry, and later extended the survey to the Lander project to the south-east. The results were streets ahead of anything existing for the Warramunga Province, and their newer products (anaglyphs and shadowgraphs created by Ian Campbell) – facilitated much better interpretation, beyond

merely the identification of ironstone anomalies;
- Austirex International went on to become a successful airborne survey contractor in Australia. It was subsequently sold to World Geoscience;
- it led to the development (by Ian Campbell) of a world-class airborne data processing software system. This was a very important asset for Austirex and later World Geoscience; and
- for many Geopeko staff it was an opportunity to work in another country, experience a different culture, and gain know-how in a different technical field.

Did it lead to the discovery of uranium in Iran? Nothing definite has ever been learned. After the maps had all been completed and sent off, Ian Campbell had a visit from two gentlemen in dark suits and dark ties. They introduced themselves as insurance agents, but later revealed that they were from ASIO and wanted the data. Ian gave them a short lesson in mineral exploration and carefully explained that the data doesn't actually confirm the presence or otherwise of a uranium deposit, and they went away empty handed.

Overall, it was a marvellous group of people involved in the project. Austirex was particularly fortunate to have four very clever people on the team – Phil McSharry, Rod Tuson, Ian Campbell, and Albert Berkavicius, without whom there would have been no show.

Others on the team from Austirex, Geosearch, Geopeko, Richardsons, and Infographics were: Tim Hooke, Tarak Malaknoodi, Pete Youngs, Rikki Youngs, Mike Pepperday, John Boyd, Ray Eaton, Robin Harbour, Dave Frazier, Bill Hay, Ken Jones, Gary Stiles, Kevin Radford, John Stewart, Merve Howel, Dave Gibson, Phil Robertson, Richard Collins, Martin Pettijohn, Peter Mewkill, Barry Harding, Peter Sceney, Harold Fuller, L Kavanis.

APPENDIX 3:
SIR JOHN PROUD

John Seymour Proud was born in Sydney on August 9th, 1907 of a well-known Sydney family, and graduated as Batchelor of Mining and Metallurgy from Sydney University in 1935. He was created Knight Batchelor in 1978, having been awarded the Institute Medal by the Australasian Institute of Mining and Metallurgy in 1974, in recognition of his services to the Australian mining industry. In addition to Fellowship of the AusIMM he was a fellow of the Institute of Engineers, a Fellow of the Institute of Marketing Management, a Life Member of the Royal Society of NSW, and a Member of the Royal Sydney Yacht Squadron.

Sir John Proud was a firm believer in the Australian mining industry. He realised that if a mining company were to survive it needed an assured source of income. He joined the board of Peko Gold Mines (Tennant Creek) NL on 26th September, 1952, and assumed the chairmanship on 28th March, 1960. His first step as Chairman was to oversee the merger of Peko with the Wallsend Holding and Investment Company Ltd, of which he was also Chairman. The rationale behind the merger was to combine a steady if not spectacular long term income from coal mining with the promise of shorter term but more spectacular bursts of higher returns from the base and metal mines that would emerge from the support of exploration from the coal mining income. From that date he guided the fortunes of the Peko-Wallsend group until his retirement in 1978, building it into one the country's premier mining houses.

Sir John believed that the best way to build a mining house was by finding your own mines, and from the day he assumed the leadership he gave unwavering, if rigorous, support to his exploration group. He encouraged and financed original research in geology and geophysics; marine biology; copper, bismuth, and titanium metallurgy; coal extraction technology; drill rig and slurry pump design; and computer software development. He was one of the first mining executives to place emphasis on environmental management by mining companies.

He was also a passionate nationalist where Australian mining was concerned. When Mt Morgan Limited was in financial trouble and attracted foreign interest in a takeover Sir John accomplished a merger with Peko-Wallsend that kept that company in Australian ownership. He then oversaw an investment program that prolonged the life of the operation for another two decades. Similarly, armed with the advice of his earth scientists, he backed their assessment that there were additional tungsten resources to be found at King Island Scheelite. Peko-Wallsend outbid a company many times its size, Goldfields of South Africa, again keeping that operation in Australian ownership.

Sir John's great asset as Chairman was his firm belief in people. His motto was "Our business is people". He was driven by a desire to build a successful mining house. He forged a successful partnership with the Electrolytic Zinc Co of Australasia in northern Australia, and using the financial strength that the mineral discoveries had delivered he expanded the group's interests into mining-related manufacturing and metal recycling. His firm belief in the expertise of his group led to the successful development and export of new Australian technologies. These included re-designed and improved rock drills and slurry pumps; a major aerial geophysical survey in Iran for which world-first electronic data collection and processing technology had been developed; innovative resource development software; and an invitation from the government of Saudi Arabia to develop a major bauxite resource there, although it later was postponed for financial reasons. The Government of PNG invited Peko-Wallsend to bid for the development of Ok Tedi when Kennecott withdrew although the bid was trumped by BHP.

In 1937 Sir John was one of two survivors from the crash of a Stinson airliner in the Lamington National Park, being kept alive for 13 days by the efforts of his fellow survivor who crawled to a nearby creek every day to bring back water in a hat, and by the fact that maggots infested his wounds and prevented them from turning gangrenous. Their rescue by Bernard O'Reilly has been the subject of a book and an ABC documentary. Sir John ensured that O'Reilly was well looked after for the rest of his life.

Sir John was a director of a range of companies and chairman of many. He was a Fellow of the Senate of Sydney University, Founder and Inaugural Chairman of the Trustees of the Lizard Island Reef Reserve Foundation, and a Trustee of the Australian Museum. He was a major supporter of Earthwatch and established the first branch of that organisation outside the USA. He also founded the Sir John Proud Fund for the Purchase of Rare Books.

Affectionately known as 'JSP' to all and sundry, Sir John's compulsory retirement at the ridiculously early age of 65 marked the start of the decline, and ultimate disappearance, of Peko-Wallsend Limited.

APPENDIX 4: VALUE OF GEOPEKO DISCOVERIES – 2015 AUSTRALIAN DOLLARS

		Found	Developed	By	Production Million AU$	Resource value Million AU$
NSW						
PARKES	NORTHPARKES	1977	1993	NORTH	$7,894	$5,213
	COWAL	1989	2005	BARRICK	$3,860	$4,480
	DUBBO ZIRCONIA			ALKANE		$38,221
COBAR	MCKINNONS	1989			$219	
QLD						
MOUNT MORGAN	MOUNT CHALMERS	1978	1980	PEKO	$190	
TOWNSVILLE	SURVEYOR	1977		KAGARA	?	$1,977
NORTHERN TERRITORY						
TENNANT CREEK	JUNO	1964	1965	PEKO	$3,265	
	IVANHOE	1961	1962	PEKO	$1,389	
	ORLANDO	1958	1960	PEKO	$1,300	
	WARREGO	1962	1969	PEKO	$10,905	
	GECKO	1969	1971	PEKO	$2,825	1,948
	ARGO	1070	1984	PEKO	$200	
	ORLANDO EAST	1985			$26	
	ROVER 1	?				$5,968
	EXPLORER 108	?				$1,878
DARWIN	RANGER 1	1970	1979	ERA	$22,166	$1,921
	RANGER 68	1976				$10,184
	BURRUNDIE?					
WESTERN AUSTRALIA						

PERTH	PEAK HILL	1980	1988	PEKO	$245	
	KANOWNA QED	1985	1990	PEKO	$61	
	KANOWNA BELLE	1990	1993	PEKO	$8,379	$424
	ABRA	1980				$10,184
TASMANIA						
KING ISLAND	DOLPHIN	1969	1973	PEKO	$583	$458
STATE TOTALS						
NEW SOUTH WALES					$11,973	$47,913
QLD					$190	$1,977
NORTHERN TERRITORY					$42,076	$21,900
WESTERN AUSTRALIA					$6,865	$9,067
TASMANIA					$583	$458

APPENDIX 5: SUMMARY OF EXPLORATION CAMPAIGNS

PROVINCE	BASE	ACQUISITION	DISCOVERY	REASON FOR PROGRAM
Warramunga	Tennant Creek	Leasing*	Black Angel Orlando Ivanho Warrego Juno Gecko Argo	Recognition by L A Richardson that the use of magnetometry would lead to the delineation of new deposits hosted in ironstone.
Kimberley	Tennant Creek	Information from prospector		Examination of Mt Angelo deposit indicated potential for the discovery of other base metal deposits.
Granites-Tanami	Tennant Creek	Leasing		The combination of a history of gold mining, rocks of similar age to those at Tennant Creek, little exploration and unexplained magnetic anomalies from the pre-WWII AGGSNA survey.
Arunta	Tennant Creek	Leasing		BMR aeromagnetics showed discrete magnetic anomalies, some coincident with copper prospector-shows suggesting magnetite-associated mineralization analogous to that at Tennant Creek.
Alligator Rivers	Jabiru	Leasing	Ranger 1 Ranger 10 Ranger 68	BMR mapping revealed Archean basement and hence analogy with Rum Jungle, with mineral deposits emplaced in near-shore traps in Lower Proterozoic.
Pine Creek (west)	Darwin Burrundie	Leasing	Mt Bonnie Quest 29 Woolwonga	Genetic model of ore formation similar to Warramunga and reinforced by existing mineral deposits, particularly Rum Jungle.

New England	Mt Morgan	Takeover	Mt Chalmers	Brownfields exploration
Tasman	King Island	Takeover	Dolphin	Brownfields exploration
Tasman	Devonport	Leasing		The established pedigree of base metal mines in the Mt Read belt in western Tasmania pointed to potential in the southern extension of the belt around Elliott Bay and at its northern end.
Tasman	Newcastle	Leasing	Tahmoor	Need to augment group coal resources.
Lachlan	Parkes	Leasing	Endeavour 7 Northparkes	Geological characteristics indicated potential for remobilisation of minerals during granite formation and geophysical patterns resembled those around established producing fields.
Lachlan	Parkes	Leasing	Lake Cowal	Targeted exploration based on geological and geophysical characteristics of the Northparkes discovery.
Lachlan	Cobar	Leasing	McKinnons Tank	To apply a new geochemical approach in the search for deposits beneath thick transported overburden.
Bangemall	Perth	JV/Leasing	Abra	Strong magnetic anomalies, considered likely to reflect deep mineral deposits, were amenable to geophysical analysis methods of drill targeting refined by L A Richardson in Tennant Creek.
Paterson	Perth	JV/Leasing		Follow up of Telfer discovery.

E Goldfields Archaean	Kalgoorlie	Leasing/JV	Kanowna Belle Mayday North Peak Hill	To assemble blocks of tenements around historically active goldfields through JV/acquisition so that broad-scaled exploration could be applied in the search for concealed deposits.
Hodgkinson	Townsville	Leasing	Surveyor 1	Follow up of genetic models responsible for the formation of 'porphyry' and 'epithermal' mineralisation.
New England	Brisbane	Leasing/JV	Rosehall	Follow up of Pajingo epithermal gold discovery.
Stuart Shelf	Adelaide	JV		Stratiform copper potential of the Stuart Shelf rocks and follow-up of the Olympic Dam discovery.
Lachlan	Ballarat	JV		Invitation to progress exploration of the East Ballarat goldfield below the depth reached by pre-WWI mining.
Stavely	Ararat	Leasing		To test the possibility that the little-known Stavely belt might contain a southward continuation of the Ordovician volcanic rocks which host the Northparkes copper-gold and Cowal gold deposits.
McArthur Basin	Mt Isa	Leasing		Re-entry into base metal exploration (IOCG and strata-bound Zn) following a corporate strategy change.
Tertiary-Quaternary sands	Brisbane and Perth	Leasing/JV	Colmer	Planned search for ancient heavy mineral deposits in the Eucla, Laura, and Perth basins (to complement proposed major acquisition of mineral sands resource by parent company North Ltd.)
Andean Copper Belt	Santiago	JV		Adjunct to assessment of advanced projects offered by Chilean government for sale by tender.

Basin and Range in Nevada	Denver	Leasing		Belief that a multi-disciplined approach to exploration with more emphasis on geophysics would bring success in search for Carlin-type gold deposits.
* 'Leasing' – exploration permits/licences and mining tenements.				

www.ingramcontent.com/pod-product-compliance
Lightning Source LLC
Chambersburg PA
CBHW071843230426
43671CB00012B/2052